Lecture Notes in Control and Information Sciences

Volume 483

This series reports new developments in the fields of control and information sciences—quickly, informally and at a high level. The type of material considered for publication includes:

1. Preliminary drafts of monographs and advanced textbooks
2. Lectures on a new field, or presenting a new angle on a classical field
3. Research reports
4. Reports of meetings, provided they are

    (a) of exceptional interest and
    (b) devoted to a specific topic. The timeliness of subject material is very important.

Indexed by EI-Compendex, SCOPUS, Ulrich's, MathSciNet, Current Index to Statistics, Current Mathematical Publications, Mathematical Reviews, IngentaConnect, MetaPress and Springerlink.

More information about this series at http://www.springer.com/series/642

Satnesh Singh · S. Janardhanan

Discrete-Time Stochastic Sliding Mode Control Using Functional Observation

Satnesh Singh (iD)
Department of Electrical Engineering
Indian Institute of Technology Delhi
New Delhi, India

S. Janardhanan
Department of Electrical Engineering
Indian Institute of Technology Delhi
New Delhi, India

ISSN 0170-8643   ISSN 1610-7411   (electronic)
Lecture Notes in Control and Information Sciences
ISBN 978-3-030-32802-3   ISBN 978-3-030-32800-9   (eBook)
https://doi.org/10.1007/978-3-030-32800-9

MATLAB is a registered trademark of The MathWorks, Inc. See mathworks.com/trademarks for a list of
additional trademarks.

This Springer imprint is published by the registered company Springer Nature Switzerland AG
The registered company address is: Gewerbestrasse 11, 6330 Cham, Switzerland

Preface

Many control applications that are encountered in practice require the design of a suitable controller to make a given system behave in a specified manner. For this purpose, the system under consideration is modelled mathematically, but a gap always lies between the original model and its mathematical model [1]. These mismatches along with the unknown external disturbances affect the system's performance. To tackle the effect of external disturbances and parameter variations, many robust control techniques have been developed. One of the most effective such techniques is variable structure control (VSC); in VSC, the structure of a closed loop is changed according to some decision rule; this rule is called the switching function [2, 3]. VSC provides a precise solution to the problem of maintaining stability and consistent performance in the face of bounded disturbances. VSC theory was first proposed in 1959 [4] and has been extensively developed since then with the invention of high-speed control devices. But due to the trouble of implementation in high-speed switching, it did not attain much popularity initially.

Sliding mode control (SMC) has been extensively recognized as a robust control strategy for its ability to make a control system very sturdy, which yields the complete elimination of external disturbances satisfying the matching conditions [5, 6]. In this control technique, a control input is designed such that the state trajectory of the system reaches a prescribed manifold in finite time and thereafter remains on it in spite of the presence of uncertainties in the system. The prescribed manifold is called a sliding manifold, the motion of the state trajectory on the sliding manifold is known as the sliding mode or sliding motion, and the corresponding control is called SMC. Since the state trajectory stays on the sliding manifold in the sliding mode, the sliding manifold alone stipulates the behaviour of the closed-loop system in the sliding mode. Therefore, the sliding manifold can be used to specify the desired performance of the system under consideration. Later, Utkin [6] presented a review paper on VSC using sliding modes that led to a renewal of interest in this area. To establish and maintain the sliding mode, control is designed such that the state trajectory is always directed towards a sliding manifold. To satisfy this constraint, SMC utilises the idea of VSC. Many practical applications of SMC have been reported in the control literature such as flight control, robotic manipulator's and servo systems [7].

Traditionally, a high-frequency switching control action is used to force the system dynamics to slide along the sliding manifold. Thus, high-frequency switching is an inherent characteristic of SMC, which results in invariance of the system state in the face of uncertainties. However, the high-frequency component in the control input leads to the problem of undesirable high-frequency vibration, called chattering, in the closed-loop system. Many methods have been developed to mitigate the effect of chattering [8, 9].

Usually, a switching type of control is used to achieve a change in controller structure. Since an increasing number of modern control systems are implemented by computers, the study in the discrete-time domain, i.e., discrete-time SMC, has been an important topic in the SMC literature. However, it has been realized that directly applying continuous-time SMC algorithms for discrete-time systems will lead to many problems, such as sample/holds effects, large chattering amplitude, discretization errors, or even instability. To cope with the aforementioned problems, the idea of discrete-time SMC (DSMC) has been introduced. Thus, the SMC design for discrete-time representation of a system is more reasonable than a continuous-time system in digitally controlled systems. As a result, there is now significant interest and research in SMC for discrete-time systems, and a number of discrete-time sliding mode (DSM) techniques have been developed [5, 10–13].

An essential property of a discrete-time system is that the control signal is computed and varied only at sampling instants, which makes discrete-time control inherently discontinuous. Hence, unlike the case of continuous SMC, the control law need not necessarily be of variable structure or have an explicit discontinuity. This method involves the design of a sliding surface that generates a stable reduced-order motion and the design of a suitable control law to force a close-loop response of a system to a sliding surface and to maintain it subsequently. Here, the system states move about the sliding manifold, but are inadequate to stay on it, whence the terminology quasi-sliding mode (QSM) [14]. In other words, the trajectories of discrete-time systems may not remain on a predesigned sliding surface because of the limitation in the sampling rate (the sampling rate cannot be infinite). However, the state trajectories may be able to remain within a boundary layer around the sliding surface called a quasi-sliding mode band (QSMB). A few studies have also proved that the chattering phenomenon in DSMC vanishes if the discontinuous control part is eliminated from the feedback control law. The control law in the absence of an explicit discontinuous component is then called a linear control law [3, 5, 11, 13]. From the above, it can be concluded that the DSMC can ensure the boundedness of the trajectories inside a QSMB, even without the use of a variable structure control strategy. This property can still be ensured in the presence of matched uncertainty in the system dynamics. From the above observations, it can be inferred that the use of a switching function in the control law may not necessarily enhance the performance [15].

Motivation

Discrete-time stochastic systems are predominant in numerous applications, and many successful attempts have been made to address the robust stabilisation of such systems [16–18]. Most of the studies in SMC available in the literature do not consider the presence of stochastic noise in the systems. However, it has been noticed that many real-world systems and a natural process may be disturbed by various noises such as process and measurement noise. This means that stochastic system representations are more aligned with reality. Therefore, it is crucial to extend the SMC theory to stochastic systems [19].

In the framework of DSMC for stochastic systems, only a few works are available in the literature [20, 21]. A method of SMC for discrete-time stochastic systems has been designed. However, in contrast with sliding function $s(k)$ design of discrete-time system and discrete-time stochastic systems is always different in nature. Sliding function design in stochastic systems is always probabilistic. Hence, the idea and definition of DSM cannot be applied directly in discrete-time stochastic systems.

The design of a controller for each control problem uses either a state feedback controller or an output feedback controller depending on the available means of measurement [22, 23]. Traditionally, SMC was developed in an environment in which all the states of the system are available. This is not a very realistic situation for practical problems and has motivated the development of functional observer-based SMC.

In this sense, this book intends to develop functional observer-based robust control strategies for stochastic systems. The use of a functional observer reduces the observer order substantially, and sliding mode control addresses the issue of robustness of the controller [24–26]. To the best of the authors' knowledge, the proposed methodology has not been previously applied to discrete-time stochastic systems. Motivated by the above observations, this book explores the problem of designing functional observer-based SMC for discrete-time stochastic systems explicitly. This book attempts to fill such gaps in the SMC literature.

The Book

The prime contributions of this book are the sliding function design for various categories of linear time-invariant (LTI) systems and its control applications, which are summarised in brief as follows:

1. SMC for discrete-time stochastic systems with bounded disturbances is designed. Subsequently, an SMC control law is designed for a stochastic system such that the states will lie within the specified band. Further, this result has been extended to the case of incomplete information, in which case, states are estimated by the Kalman filter approach and SMC is designed when the state information is not available for the systems states.

2. A functional observer-based SMC is designed for linear discrete-time stochastic systems. Sliding function, stability, and convergence analysis are given for the stochastic system. Existence conditions and stability analysis of a functional observer are provided. Finally, the controller is calculated by a functional observer method. This leads to a nonswitching type of SMC.

3. A functional observer-based SMC is designed for discrete-time stochastic systems in the presence of unmatched uncertainty. A state- and disturbance-dependent sliding function method is proposed to reduce the effect of unmatched uncertainty in the stochastic system. Finally, SMC is calculated by a functional observer method.

4. Next, DSMC is designed for parametric uncertain stochastic systems. SMC design using a functional observer is proposed for parametric uncertain discrete-time stochastic systems. SMC is calculated by a functional observer method. To mitigate the side effect of the parameters' uncertainty on the estimation of error dynamics, a sufficient condition on stability is proposed based on Gershgorin's circle theorem.

5. An SMC method is proposed for discrete-time delayed stochastic systems. Stability and convergence analysis of the proposed method are provided. Furthermore, DSMC of a delayed stochastic system for incomplete state information has also been considered, where states are estimated by the Kalman filter approach. A functional observer-based SMC method for discrete-time delayed stochastic systems is proposed. Therefore, SMC has been estimated by the functional observer approach. Finally, functional the observer-based state feedback and SMC law are compared graphically as well as numerically.

6. Next, functional observer-based SMC is developed for state time-delayed stochastic systems, in the presence of parameter uncertainties in the state and in the delayed state matrix. Finally, SMC has been calculated using a functional observer approach. To mitigate the side effect of the parameters's uncertainty on the estimation error dynamics, a sufficient condition on stability is proposed based on Gershgorin's circle theorem. A simulation example is considered to emphasize the functional observer-based SMC design.

The aim of this work is to bridge the gap between the discrete-time sliding mode and the discrete-time stochastic sliding mode by bringing in many concepts that are well defined in the former domain into the latter domain using the functional observer. It is written in a manner such that graduate students interested in sliding mode control, and particularly the discrete-time variety, will be able to grasp the difference in the design philosophy of continuous and discrete sliding modes, and we hope that it will pave the way for future research in the area of application-based discrete-time sliding mode control.

New Delhi, India   Satnesh Singh
August 2019   S. Janardhanan

Acknowledgements To our parents, teachers, and family, for the knowledge they imparted, and the support given to us, without which we would not have been capable of writing this book.

References

1. Edwards, C., Spurgeon, S.: Sliding Mode Control: Theory and Applications. Series in Systems and Control. Taylor & Francis (1998)
2. Drazenovic, B.: Automatica **5**(3), 287 (1969)
3. Gao, W., Wang, Y., Homaifa, A.: IEEE Trans. Ind. Electron. **42**(2), 117 (1995)
4. Emel'yanov, S.: IEEE Trans. Autom. Control 983–991 (1959)
5. Bartolini, G., Pisano, A., Punta, E., Usai, E.: Int. J. Control **76**(9–10), 875 (2003)
6. Utkin, V.: IEEE Trans. Autom. Control **22**(2), 212 (1977)
7. Young, K.: Variable Structure Control For Robotics and Aerospace Applications. Studies in automation and control. Elsevier (1993)
8. Acary, V., Brogliato, B., Orlov, Y.V.: IEEE Trans. Autom. Control **57**(5), 1087 (2012). 10.1109/TAC.2011.2174676
9. Acary, V., Brogliato, B., Orlov, Y.V.: IEEE Trans. Autom. Control **61**(11), 3707 (2016). 10.1109/TAC.2015.2509445
10. Tang, C.Y., Misawa, E.A.: ASME, J. Dyn. Syst. Meas. Control **122**(4), 783 (1998)
11. Bartoszewicz, A.: IEEE Trans. Industr. Electron. **45**(4), 633 (1998)
12. Furuta, K.: Syst. Control Lett. **14**(2), 145 (1990)
13. Yu, X., Wang, B., Li, X.: IEEE Trans. Industr. Inf. **8**(2), 197 (2012)
14. Ma, H., Wu, J., Xiong, Z.: IEEE Trans. Industr. Electron. **63**(10), 6292 (2016). 10.1109/TIE.2016.2580531
15. Leniewski, P., Bartoszewicz, A.: IFAC Proc Volumes **47**(3), 4589 (2014). 19th IFAC World Congress
16. Lin, H., Su, H., Shu, Z., Wu, Z.G., Xu, Y.: IEEE Trans. Autom. Control **61**(7), 1794 (2016). 10.1109/TAC.2015.2479195
17. Lu, R., Xu, Y., Zhang, R.: IEEE Trans. Industr. Electron. **63**(11), 6999 (2016). 10.1109/TIE.2016.2585543
18. Xu, Y., Lu, R., Shi, P., Li, H., Xie, S.: IEEE Trans. Cybern. **PP**(99), 1 (2017). 10.1109/TCYB.2016.2635122
19. Sharma, N.K., Singh, S., Janardhanan, S., Patil, D.U.: In: 25th Mediterranean Conference on Control and Automation, pp. 649–654 (2017)
20. Zheng, F., Cheng, M., Gao, W.B.: In [1992] Proceedings of the 31st IEEE Conference on Decision and Control, vol. 2, pp. 1830–1835 (1992). 10.1109/CDC.1992.371113
21. Zheng, F., Cheng, M., Gao, W.B.: Syst. Control Lett. **22**(3), 209 (1994)
22. Luenberger, D.: IEEE Trans. Autom. Control **11**(2), 190 (1966)
23. Luenberger, D.: IEEE Trans. Autom. Control **16**(6), 596 (1971)
24. Aldeen, M., Trinh, H.: IEE Proc. Control Theor. Appl. **146**(5), 399 (1999)
25. Darouach, M.: IEEE Trans. Autom. Control **45**(5), 940 (2000)
26. Trinh, H., Fernando, T.: Functional Observers for Dynamical Systems, vol. 420. Springer Berlin Heidelberg (2012)

Contents

Abbreviations and Symbols

Abbreviations

CSMC   Continuous-time Sliding Mode Control
LTI   Linear Time-invariant
DSM   Discrete-time Sliding Mode
SMC   Sliding Mode Control
QSM   Quasi Sliding Mode
QSMB   Quasi Sliding Mode Band
DSMC   Discrete-time Sliding Mode Control
KF   Kalman Filter
LFO   Linear Functional Observers
LQR   Linear Quadratic Regulator
LMI   Linear Matrix Inequality
SFC   State Feedback Control
SMB   Sliding Mode Band
DARE   Discrete-time Algebraic Riccati Equation
VSS   Variable Structure Systems
VSC   Variable Structure Control
VSCS   Variable Structure Control Systems
a.s.   Almost sure
p.d.s   Positive definite symmetric

Symbols

$(.)^T$ General notation for the matrix transpose operation
$\mathbb{R}$ The field of real numbers
$\mathbb{R}^n$ The n-dimensional real vector space
$\mathbb{Z}$ The set of integers

$\otimes$	Kronecker product
A	State matrix in a discrete-time system
A_d	State matrix in a discrete-time system with time delay
C	Output matrix in a discrete-time system
B	Input matrix of control in a discrete-time system
Γ	Process noise matrix in a discrete-time system
F	Unknown input matrix of the system
G	Measurement noise matrix in a discrete-time system
d	The disturbance vector effect on the sliding function
d_0	The mean of disturbance bounds d_u and d_l
d_0	Lower bound of the disturbance d
d_u	Upper bound of the disturbance d
P	Probability measure
Ω	Sample space
$\mathscr{F}$	Set of events
I_n	An $n \times n$ identity matrix
0_n	An $n \times n$ zero matrix
u	Control input
θ	Regulating the approaching speed
$\mathbb{E}(.)$	Expectation operator
$\mathbb{E}\{.|.\}$	Conditional expectation
k	Variable denoting time instant
k_x	Constant known delay in state matrix
m	Number of inputs in system
n	Number of states in system
p	Number of outputs of system
w	Vector of process noise
v	Vector of measurement noise
Q	Matrix of process noise covariance
R	Matrix of measurement noise covariance
$\|.\|$	Euclidean norm, the norm of a vector
q	Order of functional observer
s	Sliding function
$S_{\mu c}$	Sliding mode band
$\tau > 0$	Sampling time
κ, ψ	Tuning parameters
$\forall$	For all
$\square$	Designating the completion of a proof
$\rightarrow$	Tends to
$\triangleq$	Defined as
$\subset$	Proper subset of
$\cap$	Intersection
$\in$	Belongs to
$\mathbb{C}$	Complex plane

δ	Performance index
$\rho(X)$	Range space of the matrix X
x	State of a dynamical system
$\hat{x}$	Estimated system state of a dynamical system
y	System output of a dynamical system
Υ	Kalman gain
c	Sliding function parameter
K	Sliding function gain matrix
V	Lyapunov function
U	General notation for an $n \times n$ transformation matrix
L	Functional gain matrix
M, J, H, E	Unknown functional observer matrices
M_d, J_d, E_d	Unknown functional observer matrices with time delay

Chapter 1
Preliminary Concepts

Abstract Over the past several decades, a substantial amount of research work has been done in the field of SMC design and discrete-time SMC design (Veselić and Draženović in J. Frankl. Inst. 351(4):1920, 2014, Furuta in Syst. Control Lett. 14(2):145, 1990, Edwards and Spurgeon in Sliding Mode Control: Theory and Applications. Taylor & Francis, Milton Park, 1998, Janardhanan and Bandyopadhyay in IEEE Trans. Autom. Control 51(6):1030, 2006) [1–4]. In this book, we propose the design of functional observer-based SMC for discrete-time stochastic systems having deterministic, unmatched uncertain, parametric uncertain, time-delay, and parametric uncertain time delayed in nature. Therefore, a concise introduction to the idea of functional observer design for estimating the state vector and its linear function and stochastic SMC for deterministic systems is provided in this chapter.

Keywords Sliding mode control · Continuous-time systems · Observers · Reduced-order observer · State estimation · Functional observers · Process noise · Measurement noise · Stochastic sliding mode · Discrete-time systems · Kalman filter

1.1 Sliding Mode Control

Switching between two distinct control frameworks to make the resulting closed-loop system stable is called variable structure control (VSC) [5]. SMC is one of the tools available from the variable structure control systems (VSCS) toolbox. The VSCS idea has been used in the design of model-reference systems, robust regulators, tracking system, adaptive schemes, fault detection, and state observer strategies [6]. The vital aspect of SMC is a switching surface on which the system remains insensitive to internal parameter variations and external disturbance satisfying the matching condition [3].

Formally, sliding mode may be defined as follows:

© Springer Nature Switzerland AG 2020

S. Singh and S. Janardhanan, *Discrete-Time Stochastic Sliding Mode Control Using Functional Observation*, Lecture Notes in Control and Information Sciences 483, https://doi.org/10.1007/978-3-030-32800-9_1

Definition 1.1.1 (*Sliding Mode* [7]) Sliding mode or sliding motion may be defined as the evolution of the state trajectory of a system restricted to a particular nontrivial submanifold of the state space with stable dynamics.

To establish and maintain the sliding mode, the system state trajectory must converge to the sliding manifold from all the directions of the state space. If a sliding function $s(t)$ is defined, then $s(t) = 0$ represents the sliding manifold. Mathematically, this can be constructed as [3]

$$\lim_{s(t)\to 0^+} \dot{s}(t) < 0 \quad \text{and} \quad \lim_{s(t)\to 0^-} \dot{s}(t) > 0.$$

The aforementioned condition is usually written more concisely as

$$s(t)\dot{s}(t) < 0, \tag{1.1}$$

where (1.1) is referred to as the reachability condition. A succinct criterion is the η-reachability condition, which is provided as

$$s(t)\dot{s}(t) < -\eta|s(t)|,$$

where $\eta > 0$ is a positive constant. By rewriting the aforementioned equation as

$$\frac{d}{dt}s^2(t) < -\eta|s(t)|,$$

the sliding manifold becomes, as an outcome of the sliding mode, an invariant set in finite time.

1.1.1 Sliding Mode Control for Continuous-Time Systems

Consider the continuous-time LTI system

$$\dot{x}(t) = Ax(t) + Bu(t), \tag{1.2}$$

where $x(t) \in \mathbb{R}^n$ and $u(t) \in \mathbb{R}^m$ are the state and control actions, respectively. The matrices $A \in \mathbb{R}^{n\times n}$ and $B \in \mathbb{R}^{n\times m}$ are known constant matrices. The design of VSC is known to consist of the following two steps:

Step 1: The sliding function is described as

$$s(t) = cx(t), \tag{1.3}$$

where $c \in \mathbb{R}^{m \times n}$ is a constant matrix.

The sliding function is designed such that the system response is restricted to $s(t) = 0$ to achieve the desired system stability.

Step 2: The control action $u(t)$ should be of the form

$$u(x, t) = \begin{cases} u^+(x, t) & \text{if } s(x) > 0, \\ u^-(x, t) & \text{if } s(x) < 0, \end{cases} \tag{1.4}$$

where $u^+(x, t)$ and $u^-(x, t)$ are continuous functions with $u^+(x, t) \neq u^-(x, t)$.

In the physical meaning of Eq. (1.4), one designs a control action $u(x, t)$ such that any state vector $x(t)$ outside the sliding surface is the drive to reach the surface in finite time.

The equation mentioned above does not provide an adequate expression of the state velocity for $s(x) = 0$. Therefore, a procedure to resolve this issue termed the continuation method has been presented in [8].

1.1.1.1 Continuation Method

The control design considered in (1.4) defines the state velocity as

$$\dot{x}(t) = \begin{cases} f^+(x) = Ax(t) + Bu^+(t) & \text{with } s(x) > 0, \\ f^-(x) = Ax(t) + Bu^-(t) & \text{with } s(x) < 0. \end{cases} \tag{1.5}$$

The continuation method was proposed by Filippov [9] for differential equations with discontinuous right-hand sides. The methods involve constructing a solution that is an affine combination of the solution obtained by approaching the point of discontinuity from different directions. To determine the velocity of f^0 in the sliding mode, at each point on the sliding manifold, the velocities f^- and f^+ should be plotted and their ends connected [7] (Fig. 1.1).

Thus on the sliding manifold, the velocity vector is of the form

$$\dot{x}(t) = \alpha f^+ + (1 - \alpha) f^-,$$

where the scalar value $0 \leq \alpha \leq 1$ is selected such that the vector

$$f^0 \triangleq \alpha f^+ + (1 - \alpha) f^- = \dot{x}(t)$$

is tangential to the surface. The term weighting factor α is a function of x.

Calculating α using the fact that during the sliding motion one has $\dot{s}(t) = 0$ gives the state velocity as

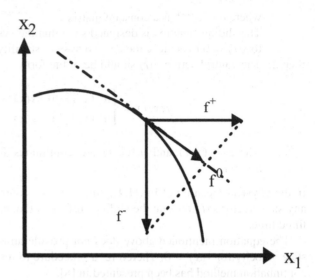

Fig. 1.1 The Filippov method to analyze the sliding mode

$$\dot{x}(t) = \frac{\dfrac{\partial s}{\partial x} f^-}{\dfrac{\partial s}{\partial x}(f^- - f^+)} f^+ - \frac{\dfrac{\partial s}{\partial x} f^+}{\dfrac{\partial s}{\partial x}(f^- - f^+)} f^-,$$

where the value of $\dot{s}(t)$ may be obtained as

$$\dot{s}(t) = \left(\frac{\partial s}{\partial t}\right) + \left(\frac{\partial s}{\partial x}\right)\dot{x}.$$

Notice that this technique can be used to determine the behavior of the plant in a sliding mode.

1.1.1.2 Equivalent SMC Design

The main intention of SMC is to ensure sliding mode in finite time from an arbitrary initial condition. The control action aims to ensure that the trajectories are driven towards and forced to remain on the sliding surface, to guarantee a sliding motion. It is usual therefore to analyze the relationship between the switching function and the control action rather than between the plant output and the control action. Suppose that at given time t_s, the switching surface is reached and an ideal sliding mode takes place. It follows that the switching function satisfies $s(x(t)) = 0, \ \forall t > t_s$, which in turn implies that $\dot{s}(x(t)) = 0, \ \forall t > t_s$. Thus the equivalent control action that maintains the sliding mode is the control action $u_{eq}(t)$ satisfying

Fig. 1.2 The reaching phase and sliding phase in sliding mode

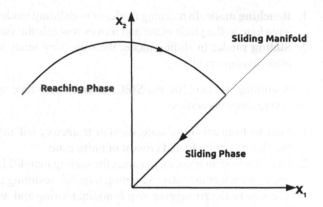

$$\dot{s}(t) = cAx(t) + cBu_{eq}(t) = 0. \tag{1.6}$$

The equivalent control can be obtained as

$$u_{eq}(t) = -(cB)^{-1}cAx(t). \tag{1.7}$$

This control law is called the equivalent control signal. The use of SMC makes the closed-loop system robust, and the closed-loop response does not depend on the variations on the system parameters, once the sliding condition is reached. Apart from the robustness properties shown by the sliding motion, another advantage arising from this situation is that the nonlinear system is forced to behave as a reduced-order system [7] (Fig. 1.2).

1.1.2 Discrete-Time Sliding Mode Control

In the last few decades, there has been an increasing interest in the control design for VSS whose structure varies under certain conditions [10, 11]. Milosavljevic [12] first introduced the main idea of DSM. He showed that in DSMC, the motion remains within some neighbourhood of the sliding surface in which the switching function is equal to zero. This sliding-like motion in discrete time is called QSM. This method involves the design of a sliding surface that generates a stable reduced-order motion and the design of a suitable control law to force a closed-loop response of the system to the sliding surface and to maintain it subsequently. Various applications of SMC have been found in areas such as robotic manipulators, aircraft, chaotic systems, process control, aerospace, motor control, and robotics [6].

The system trajectory's motion along a "chosen" surface/line/plane of the state space is called the sliding mode, which relates two modes [10]:

1. **Reaching mode**: In reaching mode, or nonsliding mode, the trajectory starts from anywhere in the phase plane and moves towards the switching line in finite time.
2. **Sliding mode**: In sliding mode, the trajectory tends to the origin of the phase plane asymptotically.

According to Gao [10], the SMC for a discrete-time system should possess the following three properties:

1. Starting from any initial state, the state trajectory will move monotonically toward the sliding manifold and cross it in finite time.
2. Once the system trajectory crosses the sliding manifold, it will cross the manifold again in each consecutive sampling interval, resulting in a zigzag motion.
3. The size of the zigzagging step is nonincreasing and within a predefined band.

1.1.2.1 State Feedback-Based DSMC

Consider the discrete-time LTI system

$$x(k+1) = Ax(k) + Bu(k). \tag{1.8}$$

The sliding function is constructed as

$$s(k) = cx(k), \tag{1.9}$$

where the state vector $x(k) \in \mathbb{R}^n$, control action $u(k) \in \mathbb{R}^m$, and sliding function $s \in \mathbb{R}^m$ are mentioned. The matrices $A \in \mathbb{R}^{n \times n}$, $B \in \mathbb{R}^{n \times m}$, and $c \in \mathbb{R}^{m \times n}$ are known. It is assumed that cB is nonsingular.

In the discrete-time implementation of the sliding mode methodology, the switching elements are replaced by a computing device that changes the structure of the system at discrete instants. It is clear that the condition (1.1), which ensures the sliding motion on the hyperplane in continuous systems, is no longer applicable in discrete-time systems. Thus, a discrete-time sliding-mode condition must be imposed. Different DSMC laws have been proposed using various reaching laws, as mentioned below:

- **Dote and Hoft law** [13]: A direct discretised version of continuous-time SMC is presented as

$$[s(k+1) - s(k)]s(k) < 0. \tag{1.10}$$

Condition (1.10) is necessary but not sufficient for the existence of a quasi-sliding motion. Indeed, it does not ensure any convergence of the state trajectories onto the hyperplane and may result in an increasing-amplitude chatter of the state trajectories around the hyperplane.

- **Sarpturk et al. law** [14]: A necessary and sufficient condition is laid down here ensuring both convergence and sliding motion onto the hyperplane. This situation may be governed as

$$|s(k+1)| < |s(k)|. \tag{1.11}$$

Condition (1.11) may, if necessary, be relaxed by also assuming equality instead of strict inequality. Here, mainly the sliding function is always directed towards the surface, and moreover, the norm of $s(k)$ decreases monotonically.

- **Gao's law** [10]: To achieve a DSMC, Gao's reaching law is adopted as

$$s(k+1) = (1 - \kappa\tau)s(k) - \psi\tau\text{sgn}(s(k)),$$

where the sampling time is $\tau > 0$, $\psi > 0$, $(1 - \kappa\tau) > 0$, $(1 - \psi\tau) > 0$, and $\kappa > 0$.

An SMC using a reaching law has been derived in [10] for the discrete-time system (1.8), and $s(k) = cx(k) = 0$ means a stable sliding surface of the form

$$u(k) = -(cB)^{-1}((cA - c + \kappa\tau c)x(k) + \psi\tau\text{sgn}(s(k))). \tag{1.12}$$

However, this control law would yield only a QSM and would also introduce a chattering of amplitude

$$\delta = (2I - \kappa\tau)^{-1}\psi\tau.$$

The control law used in (1.12) has two parameters, κ and ψ, for tuning the response.

- **Linear reaching law**: In order to achieve the condition (1.11), a modified reaching law has been proposed by Hui and Zak [15] and Ming [16]:

$$s(k+1) = \theta s(k), \tag{1.13}$$

where the range of θ is $0 < \theta < 1$. The reaching law (1.13) is very common to Gao's and Sarpturk's reaching laws. However, the reaching law (1.13) gives a perfect explanation of the appropriate trajectory toward the sliding surface. DSMC algorithms mainly try to satisfy the condition

$$s(k+i) = 0 \quad \text{for some} \ i \geq 1.$$

The resultant control law, in the case of $i = 1$, using the algorithm proposed in [7], will be the form

$$u(k) = -(cB)^{-1}cAx(k). \tag{1.14}$$

Note that the control action is no longer of variable structure. However, it is an SMC technique. Hence, it can be seen that through an SMC initiated from VSC, during the process of its advancement, the idea of a sliding mode became free of variable structure and switching.

1.1.3   Discrete-Time Sliding Mode Control for Stochastic
Systems

A stochastic system is a static system whose processes are characterised by probability distributions. Recently, to include the uncertainties in the mathematical model,
probabilistic methods have been developed. Unlike deterministic partial differential
equations, solutions of stochastic partial differential equations are random fields,
which results in a challenging problem for nonlinear systems with a large number of
stochastic parameters.

A filtered probability space $(\Omega, \mathscr{F}, \mathscr{F}_{k\geq 0}, P)$ is described as follows [17]:

- **Sample space** (Ω): This space may be defined as a set that comprises every
  possible outcome, and their combinations form a random experiment/process.
- **Set of events** $(\mathscr{F})$: This set is a subset of the sample space (Ω). An event set is
  a collection of one outcome or a combination of possible outcomes. (This set can
  also be the null set, representing no outcome.)
- **Filtration** $(\mathscr{F}_{k\geq 0})$: Filtrations are used to model the information that is available
  at a given point and therefore play an influential role in the formation of random
  processes.
- **Probability measure** (P): A probability measure represents the probability of a
  possible outcome/combination of outcomes. It is basically a mapping between the
  event set $(\mathscr{F})$ and a set whose elements lie in the set $(0, 1)$, i.e., $P : \mathscr{F} \rightarrow [0, 1]$.

1.1.3.1   Discrete-Time Sliding Mode Control for Stochastic Systems

Consider an LTI discrete-time stochastic system

$$x(k+1) = Ax(k) + Bu(k) + \Gamma w(k), \tag{1.15}$$

where the matrices $A \in \mathbb{R}^{a\times n}$, $B \in \mathbb{R}^{n\times m}$, and $\Gamma \in \mathbb{R}^{n\times \tilde{r}}$ are known. The plant noise
$w(k) \in \mathbb{R}^{\tilde{r}}$ is of mean zero and the Gaussian white noise processes have covariance Q.

As defined in (1.9), the sliding function is given by

$$s(k) = cx(k). \tag{1.16}$$

Assumption 1.1 cB is a nonsingular matrix.

The main goal is to find an SMC $u(k)$ such that the system states (1.15) will be
driven to and held within the following band in $\mathbb{R}^n$ [18]:

$$S_{\mu c} = \{x \in \mathbb{R}^n :| cx | \leqslant \mu_c\}, \quad \mu_c > 0, \tag{1.17}$$

with probability $(1 - \delta)$, i.e.,

$$P\{|s(k)| \leqslant \mu_c\} = (1 - \delta) \quad \text{for} \quad k > N, \tag{1.18}$$

where $0 < \delta \ll 1$ and $N \in \mathbb{Z}$ is sufficiently large. Here $S_{\mu c}$ is called the sliding mode band (SMB).

Remark 1.1 In traditional DSMC systems, it is usually assumed that the system signals can be successfully transmitted to the controller or actuator. However, in a probabilistic sense, states will lie within a band with probability almost surely. Therefore, DSMC in a probabilistic sense is suitable for the stochastic DSMC problem. In fact, no control methodology dealing with a stochastic system can define with surety that the performance is within a band. Only a high probability can be ensured.

Proposition 1.1 ([18]) *The objective* (1.18) *can be ensured by the following reaching condition:*

$$P\{|s(k+1)| < \theta|s(k)|||\mathscr{F}_k\} \geq 1 - \varepsilon \quad a.s. \tag{1.19}$$

where $0 < \theta < 1, 0 < \varepsilon < 0.5$, *and* $\mathscr{F}_k \subset \mathscr{F}$ *denotes the* σ*-field defined by* $\{x(k), x(k-1), \dots x(0); u(k), u(k-1), \dots u(0)\}$. *The parameter* θ *is admitted in* (1.19) *for regulating the approaching speed.*

Proof On considering $m_{sa}(k+1) = \mathbb{E}\{|s(k+1)|\}$
and $m_{sa}(k+1|k) = \mathbb{E}\{|s(k+1)|||\mathscr{F}_k\}$, we claim that

$$m_{sa}(k+1|k) \leqslant \theta|s(k)| \quad a.s. \tag{1.20}$$

is ensured by (1.19). Suppose, for the sake of a contradiction, that there exists a set $D \subset \Omega$ with $P(D) > 0$ such that $m_{sa}(k+1|k) > \theta|s(k)|, \forall \omega \in D$. It follows that

$$P\{|s(k+1)| \geq \theta|s(k)|||\mathscr{F}_k\} \geq P\{|s(k+1)| \geq m_{sa}(k+1)|\mathscr{F}_k\} = 0.5 \quad \forall \omega \in D. \tag{1.21}$$

On the other hand, it follows from (1.19) that

$$P\{|s(k+1)| \geq \theta|s(k)|||\mathscr{F}_k\} \geqslant 1 - (1 - \varepsilon) = \varepsilon \quad \forall \omega \in D_1, \tag{1.22}$$

where $D_1 \subset \Omega$ and $P(D_1) = 1$. Clearly, $P(D_1 \cap D) > 0$. This, along with (1.21) and (1.22), yields that $\varepsilon \geq 0.5$, which contradicts the fact that $\varepsilon < 0.5$. Hence the claim (1.20) holds.

Applying the expectation operator to both sides of (1.20) yields

$$m_{sa}(k+1) = \mathbb{E}(m_{sa}(k+1|k)) \leqslant \mathbb{E}(\theta|s(k)|) = \theta m_{sa}(k). \tag{1.23}$$

Since $m_{sa}(k) \geq 0$, it follows from (1.23) that $\lim_{k \to \infty} m_{sa}(k) = 0$.

By the Markov inequality [19], we have

$$P\{|s(k)| \geq \mu_c\} \leq \frac{\mathbb{E}(|s(k)|)}{\mu_c} = \frac{m_{sa}(k)}{\mu_c}, \tag{1.24}$$

and therefore

$$P\{|s(k)| \leq \mu_c\} \leq 1 - \frac{m_{sa}(k)}{\mu_c}. \tag{1.25}$$

So (1.18) holds for suitably chosen real numbers μ_c and δ if N is large enough.

Lemma 1.1 *Consider a Gaussian random variable g with mean zero and variance σ_g^2, and let W_g be the given solution of the equation*

$$P\{|g| \leq W_g\} = \mathscr{L}(W_g, -W_g, 0, \sigma_g) = 1 - \varepsilon. \tag{1.26}$$

Let $g_\rho = \rho + g$. Then the solutions of the equation

$$P\{|g_\rho| \leq \phi\} = \mathscr{L}(\phi, -\phi, \rho, \sigma_g) = 1 - \varepsilon \tag{1.27}$$

with respect to ρ have following properties:

  (i) *If $\phi > W_g$, then (1.27) has only two solutions.*
 (ii) *If $\phi > W_g$ and $0 < \varepsilon < 0.5$, then the solutions of (1.27) are bounded by ϕ, i.e., $|\rho| \leq \phi$.*

Proof For simplicity, the proof of Lemma 1.1 is not given here; please refer to [18] for a detailed proof.

1.1.3.2   Synthesis of DSMC

Here our main goal is find the control that achieves the objective as in (1.18) for the system (1.15).

Theorem 1.2 *There occur predictable values of W_c, $m_1^{\pm}(k)$, and $m_2^{\pm}(k)$ such that the DSMC law*

1. *$u(k) = u^+(k) \in (cB)^{-1}(-cAx(k) + m_1^+(k), -cAx(k) + m_2^+(k))$ if $s(k) > \theta^{-1}W_c$,*
2. *$u(k) = u^-(k) \in (cB)^{-1}(-cAx(k) + m_1^-(k), -cAx(k) + m_2^-(k))$ if $s(k) < -\theta^{-1}W_c$,*
3. *$u(k) = -(cB)^{-1}cAx(k)$ if $|s(k)| \leq \theta^{-1}W_c$,*

drives the system states (1.15) to the SMB $S_{\mu c}$ and held within it with a given probability $(1 - \delta)$.

Proof Consider $s(k + 1) = m(k + 1) + g(k + 1)$.
      Then

$$m(k + 1) = cAx(k) + cBu(k)$$

and

$$g(k+1) = c\Gamma w(k),$$

where $g(k+1)$ is Gaussian with mean zero and variance $\sigma_c^2 = c\Gamma Q\Gamma^T c^T$.

Case 1.1 $s(k) > \theta^{-1}W_c$.

In this case, (1.19) can be brought into the form

$$P\{-\theta s(k) < s(k+1) < \theta s(k)\} \geq 1 - \varepsilon. \tag{1.28}$$

A condition necessary and sufficient for the survival of solutions of (1.28) is

$$s(k) > \theta^{-1}W_c, \tag{1.29}$$

where W_c is represented by

$$\mathcal{L}(W_c, -W_c, 0, \sigma_c) \triangleq \int_{-W_c}^{W_c} \frac{1}{\sqrt{2\pi\sigma_c^2}}e^{-z^2/2\sigma_c^2}dz = 1 - \varepsilon, \tag{1.30}$$

for a Gaussian random variable with mean zero and variance σ_c^2.

On solving Eq. (1.30), we obtain

$$W_c = \sqrt{2}\sigma \ (\text{erf})^{-1}(1 - \varepsilon), \tag{1.31}$$

where erf is the error function [20].

By Lemma 1.1, under the condition (1.29), the controller is

$$u(k) = u^+(k) \in (cB)^{-1}(-cAx(k) + m_1^+(k), -cAx(k) + m_2^+(k)), \tag{1.32}$$

where $u \in [\beta_1, \beta_2]$ means that any number in the closed interval $[\beta_1, \beta_2]$ can be selected as the control action value u, and $m_i^+(k), i = 1, 2$, are the two solutions of the following equation satisfying $m_1^+(k) < 0 < m_2^+(k)$:

$$\mathcal{L}(\theta s(k), -\theta s(k), \rho, \sigma_c) = 1 - \varepsilon. \tag{1.33}$$

Case 1.2 $s(k) < -\theta^{-1}W_c$.

In this case, (1.19) is of the form

$$P\{\theta s(k) < s(k+1) < -\theta s(k)\} \geq 1 - \varepsilon. \tag{1.34}$$

Similarly, a condition that is necessary and sufficient for the survival of solutions of (1.34) is

$$s(k) < -\theta^{-1}W_c, \tag{1.35}$$

and the controller will be

$$u(k) = u^-(k) \in (cB)^{-1}(-cAx(k) + m_1^-(k), -cAx(k) + m_2^-(k)), \qquad (1.36)$$

where $m_i^-(k)$, $i = 1, 2$, are the only two solutions of the following equation in ρ that satisfy $m_1^-(k) < 0 < m_2^-(k)$:

$$\mathscr{L}(-\theta s(k), \theta s(k), \rho, \sigma_c) = 1 - \varepsilon. \qquad (1.37)$$

Case 1.3 $|s(k)| \leq \theta^{-1}W_c$.

In this case, the SMC should be selected as

$$u(k) = -(cB)^{-1}cAx(k). \qquad (1.38)$$

From Lemma 1.1, it can be seen that the control input (1.38) will be within the valid range of values for $u(k)$. Hence it can be used for the SMC implementation. It is also valid for Cases 1 and 2.

Remark 1.2 When the SMC (1.38) is used, it will be

$$s(k + 1) = c\Gamma w(k). \qquad (1.39)$$

In this case, the system motion (1.15) is called the stochastic sliding mode (SSM).

Remark 1.3 In a more general case, smaller values of ε and μ_c or W_c improve the performance of the controlled system. But this requirement for ε and μ_c is contradictory, because W_c is determined by ε according to (1.31) and conversely, and it is clear that the smaller the value of ε, the larger that of W_c for a given $\sigma_c > 0$. Precisely, $\varepsilon = 0$ and $\mu_c = 0$, which hold for deterministic SMC systems if and only if $\sigma_c = 0$, i.e., the plant noise satisfies $w(k) = 0$ almost surely, and hence $w(k)$ can be treated as a deterministic process. In this case, as expected, the stochastic SMC reduces to deterministic SMC.

Simulation Example

Consider the discrete-time stochastic system (1.15) with parameters

$$A = \begin{bmatrix} 1 & 1 \\ 0 & 1 \end{bmatrix}, B = \begin{bmatrix} 0.5 \\ 1 \end{bmatrix}, \Gamma = \begin{bmatrix} 1 & 0.5 \\ 0 & 1 \end{bmatrix}, c = \begin{bmatrix} 1 & 1 \end{bmatrix}.$$

The initial state conditions are given by $x(0) = [10 \ \ 10]$, where $w(.)$ is a white Gaussian noise with zero mean and variance 0.25. The parameters θ and ε are chosen as $\theta = 0.9$ and $\varepsilon = 0.10$.

Figure 1.3a, b are the state trajectories, Fig. 1.3c is the control input, and Fig. 1.3d is the sliding function within a specific band.

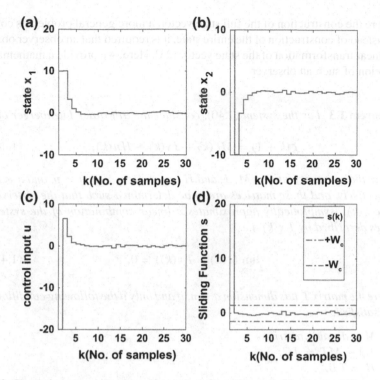

Fig. 1.3 Simulation responses of complete state information: **a** and **b** state variables, **c** control input, and **d** sliding function

1.2 Observers

A classical method for estimation problem in deterministic systems is the Luenberger observer. In most control problems, for example pole placement, dead beat control and linear quadratic regulator design require system states for their implementation. The control law for the problems mentioned above is just a linear functional of the state vector if the general full state vector is not available. There is a need of an auxiliary system for the solution of the incomplete state information and the reconstruction of the state vector. The theory of state observers evolved by the pioneering work by D. G. Luenberger (1963) and is now well established. An observer is proposed for the excitation control of a micro-alternator to implement linear feedback control for which inaccessible states are reconstructed in [21].

Luenberger [22] first investigated the problem of observing the system state of a deterministic LTI. Consider a discrete-time LTI system represented by

$$
\begin{aligned}
x(k+1) &= Ax(k) + Bu(k), \\
y(k) &= Cx(k).
\end{aligned}
\tag{1.40}
$$

Before the construction of the full state vector, a more general problem is considered. Instead of construction of the entire state, it is required that an observer observe some linear transformation of the state vector [23]. Here, we provide a mathematical description of such an observer.

Theorem 1.3 *For the system* (1.40), *consider the qth-order Luenberger observer*

$$\zeta(k+1) = M\zeta(k) + Jy(k) + Hu(k), \qquad (1.41)$$

where the constant matrices M, J, and H are of size $q \times q$, $q \times p$, and $q \times m$ respectively, and these matrices are to be determined such that the observer state $\zeta(k)$ asymptotically approximates a linear combination of the system states described by $Tx(k)$, i.e.,

$$\lim_{k\to\infty} (\zeta(k) - Tx(k)) = 0, \qquad (1.42)$$

where the matrix T has dimension $q \times n$, if and only if the following conditions are satisfied:

(i) *M is a stable matrix;*
(ii) *$TA - MT = JC$;*
(ii) *$H = TB$.*

Proof The estimation error is described as

$$e(k) = \zeta(k) - Tx(k). \qquad (1.43)$$

From (1.40) to (1.41), we obtain the following equation for the error dynamics:

$$e(k+1) = Me(k) + (JC - TA)x(k) + (H - TB)u(k). \qquad (1.44)$$

If there exists a matrix T such that conditions (ii) and (iii) of Theorem 1.3 are satisfied, then the error observer dynamics (1.44) becomes $e(k+1) = Me(k)$. In addition, if (i) is also satisfied, it is seen that the error $e(k)$ will asymptotically approach zero for any $x(0)$ and $\zeta(0)$. Hence the observer state vector $\zeta(k)$ is an asymptotic estimate of $Tx(k)$.

1.2.1 State Observation

Now we consider the problem of observing the full state vector. By forcing the transformation matrix T to a unit matrix of order n, i.e., $T = I_n$, then (1.40) will be

$$\hat{x}(k+1) = (A - JC)\hat{x}(k) + Jy(k) + Hu(k), \tag{1.45}$$

where $\hat{x}(k) = \zeta(k)$ is an estimate of $x(k)$, and its dimension is the same as that of the original system, i.e., $q = n$. This auxiliary system is known as a full order or an identity order. A state observer is also called a state estimator.

The dynamics of the full-order observer is resolved by the eigenvalues of $(A - JC)$. If J is chosen in such a manner that the eigenvalues of $(A - JC)$ are at appropriate locations, then the observer can be made to have desired dynamics.

Proposition 1.4 *All eigenvalues of $(A - JC)$ can be assigned arbitrarily by selecting a real constant matrix G if and only if the system (1.40) is completely observable. As a notational convenience, this situation will sometimes be described by writing "(C, A) is completely observable."*

However, the above method of state estimation possesses a measure of redundancy. Since p-outputs are linear combinations of the state variables available through system output, $y(k) = Cx(k)$, it might seem that we could attempt to estimate only $(n - p)$ other (independent) linear combination of the state vector, in order to reconstruct the full state. This expectation is indeed fulfilled in [22]. Such observers are known as reduced-order or minimum-order state observers. Then an estimate $\hat{x}(k)$ of $x(k)$ can be determined through

$$\hat{x}(k) = \begin{bmatrix} T \\ C \end{bmatrix}^{-1} \begin{bmatrix} \zeta(k) \\ y(k) \end{bmatrix}, \tag{1.46}$$

where $\zeta(k) \in \mathbb{R}^{n-p}$ is the state of the observer that estimate $Tx(k)$ and the $n \times n$ matrix $\begin{bmatrix} T \\ C \end{bmatrix}$ is nonsingular.

1.2.2 Observer-Based SMC for Discrete-Time Systems

The states of the system are not usually accessible, or there may exist unavoidable noise in the measured states. These problems may affect the performance of the system. Therefore, utilising observers can resolve such issues. The design of the controller for each control problem uses either SFC or an output feedback-based controller depending upon the availability of measurement [7, 24, 25]. The issue of states estimation of linear systems is of great importance in many applications, because the states are not available in many systems and therefore must be estimated.

1.2.3 Observer-Based SMC for Discrete-Time Stochastic Systems

Luenberger observers are very simple and effective. They have many problems in dealing with random noise or certain inputs, so they do not operate properly in estimating the states. The impetus for the development of observers for linear multivariable systems was given by Kalman (1960) when he introduced state space realisation for LTI systems [26]. The Kalman filter (KF) has solved the estimation problem in random linear systems as a linear filter by use of the minimum mean square error [26, 27]. In using KF, it is supposed that the system parameters, covariance of the process, and calculation noises and also system inputs are all known [19, 20, 28]. The problem of state estimation for discrete-time stochastic complex network has been proposed [29–33]. A robust observer-based DSMC is proposed for networked control systems, including random pocket losses [34–36]. In the context of stochastic SMC, few works have been available in the literature in recent years [37].

The reaching condition would be different in the case that states are not fully accessible. The output equation is described as

$$y(k) = Cx(k) + Gv(k), \tag{1.47}$$

where $v(k) \in \mathbb{R}^l$ is white Gaussian noise with mean zero and covariance R. The matrix $G \in \mathbb{R}^{p \times l}$ is known.

The sliding function is described as

$$\hat{s}(k) \triangleq c\hat{x}(k), \tag{1.48}$$

where $\hat{x}(k)$ denotes the state estimation of the original state $x(k)$ given $\tilde{\mathscr{F}}_k$, the σ-field created by $\{y(k), y(k-1), \ldots y(0); u(k), u(k-1), \ldots u(0)\}$.

The main task is to control the sliding variable $\hat{s}(k)$ such that

$$P\{|\hat{s}(k)| \leqslant \tilde{\mu}_c\} = (1 - \delta) \quad \text{for} \quad k > N, \tag{1.49}$$

where $\tilde{\mu}(k) > 0$, and then analyse the behaviour of $s(k)$.

1.2.3.1 An Algorithm for State Estimation

1. $\hat{x}(0) = x_0$ is initially assumed.
2. Given the noise covariance $P(k-1)$ of the state vector $x(k-1)$ and the covariances of $w(.)$ and $v(.)$ as Q and R, the error covariance $P(k)$ is determined as [28]

$$(a)\ P(k|k-1) = AP(k-1)A^T + \Gamma Q \Gamma^T,$$
$$(b)\ P(k) = [P^{-1}(k|k-1) + C^T R^{-1} C]^{-1}, \tag{1.50}$$
$$(c)\ \Upsilon(k) = P(k)C^T R^{-1},$$

where $\Upsilon(k)$ is the Kalman gain.

3. Using $\Upsilon(k)$, the state vector $\hat{x}(k)$ can be determined as

$$(a)\ \hat{x}(k|k-1) = A\hat{x}(k-1) + Bu(k-1),$$
$$(b)\ \hat{x}(k) = \hat{x}(k|k-1) + \Upsilon(k)[y(k) - C\hat{x}(k|k-1)]. \tag{1.51}$$

1.2.3.2 Controller Design

Using Proposition 1.19, (1.49) can be obtained by the reaching condition

$$P\{|\hat{s}(k+1)| < \theta|\hat{s}(k)|\,|\tilde{\mathscr{F}}_k\} \geq 1 - \varepsilon \quad a.s. \tag{1.52}$$

The estimation error between the original and estimated states is described as $\tilde{x}(k) \triangleq x(k) - \hat{x}(k)$.

Theorem 1.5 *There occur predictable values of* $\tilde{W}(k+1), m_1^{\pm}(k), m_2^{\pm}(k)$ *such that the DSMC law*

1. $u(k) = u^+(k) \in (cB)^{-1}(-cA\hat{x}(k) + \tilde{m}_1^+(k), -cA\hat{x}(k) + \tilde{m}_2^+(k)])\ if\ s(k) > \theta^{-1}\tilde{W}(k+1),$
2. $u(k) = u^-(k) \in (cB)^{-1}(-cA\hat{x}(k) + \tilde{m}_1^-(k), -cA\hat{x}(k) + \tilde{m}_2^-(k))\ if\ s(k) < -\theta^{-1}\tilde{W}(k+1),$
3. $u(k) = -(cB)^{-1}cA\hat{x}(k)\ if\ |s(k)| \leq \theta^{-1}\tilde{W}(k+1),$

drives the system states (1.15) to the band $S_{\mu c}$ *and held within it with the given probability* $(1 - \delta)$. *The solution of* $\tilde{W}(k+1)$ *is defined by*

$$\mathscr{L}(\tilde{W}(k+1), -\tilde{W}(k+1), 0, \tilde{\sigma}(k+1)) \triangleq \int_{-\tilde{W}(k+1)}^{\tilde{W}(k+1)} \frac{1}{\sqrt{2\pi\tilde{\sigma}^2(k+1)}} e^{-(z-0)^2/2\tilde{\sigma}^2(k+1)} dz$$
$$= 1 - \varepsilon, \tag{1.53}$$

with a Gaussian random variable with zero mean and variance $\tilde{\sigma}^2(k+1)$.

Now the SMC law is designed as follows:

Proof On considering

$$\tilde{m}(k+1) \triangleq cA\hat{x}(k) + cBu(k) \tag{1.54}$$

and
$$\tilde{g}(k+1) \triangleq c\Upsilon(k+1)[CA\tilde{x}(k) + C\Gamma w(k) + Gv(k+1)]. \tag{1.55}$$

Applying (1.51) directly to (1.48) yields

$$\hat{s}(k+1) \triangleq \tilde{m}(k+1) + \tilde{g}(k+1). \tag{1.56}$$

Here $\tilde{g}(k+1)$ is a Gaussian random variable with mean zero and variance

$$\tilde{\sigma}^2(k+1) = c\Upsilon(k+1)[C(AP(k)A^T + \Gamma Q\Gamma^T)C^T + R)]\Upsilon^T(k+1)c^T. \tag{1.57}$$

By (1.50), it follows that

$$\tilde{\sigma}^2(k+1) = c[P(k+1|k) - P(k+1)]c^T. \tag{1.58}$$

Case 1.4 If $\hat{s}(k) > \theta^{-1}\tilde{W}(k+1)$, then the SMC is

$$u(k) = u^+(k) \in (cB)^{-1}(-cA\hat{x}(k) + \tilde{m}_1^+(k), -cA\hat{x}(k) + \tilde{m}_2^+(k)), \tag{1.59}$$

where $\tilde{m}_i^+(k)$, $i = 1, 2$, are the two solutions of the following equation satisfying $\tilde{m}_1^+(k) < 0 < \tilde{m}_2^+(k)$:

$$\mathscr{L}(\theta\hat{s}(k), -\theta\hat{s}(k), \tilde{\rho}, \tilde{\sigma}(k+1)) = 1 - \varepsilon.$$

Case 1.5 If $\hat{s}(k) < -\theta^{-1}\tilde{W}(k+1)$, then the SMC

$$u(k) = u^-(k) \in (cB)^{-1}(-cA\hat{x}(k) + \tilde{m}_1^-(k), -cA\hat{x}(k) + \tilde{m}_2^-(k)), \tag{1.60}$$

where $\tilde{m}_i^-(k)$, $i = 1, 2$, are the only two solutions of the following equation concerning $\tilde{\rho}$ that satisfy $\tilde{m}_1^-(k) < 0 < \tilde{m}_2^-(k)$:

$$\mathscr{L}(-\theta\hat{s}(k), \theta\hat{s}(k), \tilde{\rho}, \tilde{\sigma}(k+1)) = 1 - \varepsilon.$$

Case 1.6 If $|\hat{s}(k)| \leq \theta^{-1}\tilde{W}(k+1)$, then SMC should be obtained:

$$u(k) = -(cB)^{-1}cA\hat{x}(k). \tag{1.61}$$

From Lemma 1.1, it can be seen that the control input (1.61) will be within the valid range of values for $u(k)$. Hence, it can be used for the SMC implementation. It is also valid for Cases 1 and 2.

Remark 1.4 When the SMC (1.61) is used, the sliding variable $s(k)$ will have the dynamics

$$\hat{s}(k+1) = c\Upsilon(k+1)[CA\tilde{x}(k) + C\Gamma w(k) + v(k+1)]. \tag{1.62}$$

From (1.56) and (1.62), it can be seen that the conditional variance of $\hat{s}(k + 1)$ will be minimal when the system reaches the stochastic sliding mode.

1.2.3.3 Simulation Results

Consider the discrete-time LTI stochastic system (1.47):

$$y(k) = [1 \quad 0]x(k) + v(k)$$

and

$$s(k) = [1 \quad 1]x(k),$$

where $v(.)$ is white Gaussian noise with mean zero and variance 1.0. The observer states' initial values are $\hat{x}(0) = [-15 \quad 0]$ and $P(0) = 4.0I_2$. The parameters θ and ε are chosen as $\theta = 0.9$ and $\varepsilon = 0.10$.

In case of incomplete state information, Fig. 1.4a, b are the estimated state trajectories, Fig. 1.4c is the estimated control input, and Fig. 1.4d is the estimated sliding function within a specific band.

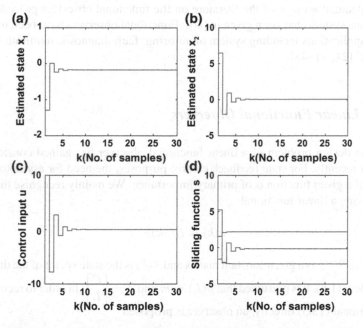

Fig. 1.4 Simulation responses of incomplete state information: **a** and **b** state variables, **c** control input, and **d** sliding function

1.3 Functional Observers

The basic issue with the state observers is a substantial increase of the overall dynamics. The state observer is an auxiliary system added to the original system. The poles of the auxiliary system are added to the pole of the original open-loop system in a closed-loop operation. This will increase the complexity of the hardware as well as the software simulation of the overall closed-loop system. Poles of the auxiliary system also contribute to the error dynamics of the overall closed-loop system, and the estimation error will propagate through the original system. These issues are the main motivation of the present work on functional observers. The functional observers give a remedy for the above-mentioned problems because of their reduced dynamics compared to those of the state observers for multivariable linear systems, which is why the design of low-dimensional functional observers is an alluring topic of research.

The concept of the functional observer was first introduced in [22] by Luenberger. The estimation of a function of states does not necessarily require the estimation of all the states, and therefore the order of the functional observers can be relatively less than that of the full state observer. The functional observer concept has been extended to the problem of multilinear systems [38]. A necessary and sufficient condition for the design and existence of functional observers was proposed by Darouach [39]. A detailed account of the literature on the functional observers procedure for continuous systems has been given in [40]. Functional observers have been used for several applications including system monitoring, fault diagnosis, multirate output, and SMC [24, 41–48].

1.3.1 Linear Functional Observers

The main design problem of a linear functional observer has gained considerable attention recently. For state feedback control purposes, the need for asymptotic observers of a given function is of primary importance. We mainly recognise the issue of observing a linear functional

$$u(k) = Lx(k), \tag{1.63}$$

where $L \in \mathbb{R}^{m \times n}$ is a given constant matrix and $x(k)$ is the state vector of the discrete-time system (1.8). It is presumed that $\rho(L) = m$ and $\rho = \begin{bmatrix} C \\ L \end{bmatrix}$. In order to reconstruct the functional (1.63) directly, an observer is proposed

$$\begin{aligned} \zeta(k+1) &= M\zeta(k) + Jy(k) + Hu(k), \\ \hat{u}_f(k) &= V\zeta(k) + Ey(k), \end{aligned} \tag{1.64}$$

where $\zeta(k) \in \mathbb{R}^q$ is the observer state vector, and $\hat{u}_f(k)$ is the estimate of $x(k)$, the functional state vector. The matrices $M \in \mathbb{R}^{q \times q}$, $J \in \mathbb{R}^{q \times p}$, $H \in \mathbb{R}^{q \times m}$, $V \in \mathbb{R}^{r \times q}$, and $E \in \mathbb{R}^{r \times p}$ are unknown constants.

Theorem 1.6 *The qth-order observer* (1.64) *will estimate* $u_f(k)$ *if the following conditions hold:*

(i) *M is a stable matrix,*
(ii) *$VT = L - EC$,*
(iii) *$MT + JC - TA = 0$,*
(iv) *$H = TB$,*

where $L \in \mathbb{R}^{r \times n}$ is the functional gain matrix and $T \in \mathbb{R}^{q \times n}$ is an unknown constant matrix.

Proof Let us define $e(k)$, the error between $\zeta(k)$ and $Tx(k)$, as

$$e(k) \triangleq \zeta(k) - Tx(k). \tag{1.65}$$

Taking the first-order difference of (1.65) gives

$$
\begin{aligned}
e(k+1) &= \zeta(k+1) - Tx(k+1) \\
&= M\zeta(k) + Jy(k) + Hu(k) - TAx(k) - TBu(k) \\
&= Me(k) + (MT + JC - TA)x(k) + (H - TB)u(k).
\end{aligned}
\tag{1.66}
$$

If conditions (iii)–(iv) are satisfied, then the error dynamics (1.66) will be

$$e(k+1) = Me(k). \tag{1.67}$$

Hence, the error dynamics $e(k)$ is governed by the matrix M.

The output error is constructed as

$$
\begin{aligned}
e_u(k) &\triangleq \hat{u}_f(k) - Lx(k) \\
&= V\zeta(k) + Ey(k) - Lx(k) \\
&= Ve(k) + (VT + EC - L)x(k).
\end{aligned}
\tag{1.68}
$$

If condition (ii) is satisfied, then the output error dynamics (1.68) will be

$$e_u(k) = Ve(k). \tag{1.69}$$

Hence, the output error dynamics (1.69) is governed by the matrix V. This completes the proof.

From the above derivation, it is clear that the functional $z(k)$ can be computed using the state feedback. The computed functional can be just a functional or can be used directly as control input in a feedback system. In some cases, the computed functional can also be used for the computation of a control input functional such as SMC. This technique can also be used for functional computation even if the system is not observable. It is clear that the proposed work on functional computation generalises the solution, since it includes the solution of the set of unobservable LTI systems. It is evident by the functional observer method that the number of outputs required is less than in other techniques, and this will reduce the computational burden for obtaining a static structure of the observer. Here, the functional observer is used in a closed-loop system for estimation of the control sequence, but in general, it can be used in an open-loop system for the estimation of a linear function of the state vector.

Following the contributions of Luenberger [22], various algorithms have been proposed in the literature for the functional observer design. A complete survey of existing algorithms for the design of functional observers is given in [40] and future directions is also given. These algorithms have been classified primarily based on the number of functions to be estimated.

1.3.2 Main Advantages of Functional Observers

Functional observers are a recurring theme in state feedback control. That is, in feedback, a function, even only a linear combination of the state variables $Lx(k)$, is generally required, rather than the entire knowledge of the state vector [40]. Direct estimation will reduce the order of the observer considerably in comparison with the state observer. From a practical point of view, designing a feasible order of observer will directly affect the components, leading an improved cost–benefit ratio, which in turn increases reliability. When this functional observer exists, its order, namely q according to our notation, is less than the order $(n - p)$ of a reduced-order state observer. From the above observations, it can be inferred that full-order observers, reduced-order observers, and partial-state observers can all be considered under one unified framework.

According to the q-dimensional state observer $\zeta(k)$, several observers can be distinguished (Table 1.1).

Table 1.1 Distinguished observer and order of observers

Observer	Order of observer (q)
Full-order [22]	$q = n$
Reduced-order [49]	$q = (n - p)$
Functional [39]	$q \leq (n - p)$
Minimum-order [50]	$q = m$

References

1. Veselić, B., Draženović, B.: Čedomir Milosavljević. J. Frankl. Inst. **351**(4), 1920 (2014). (Special Issue on 2010-2012 Advances in Variable Structure Systems and Sliding Mode Algorithms)
2. Furuta, K.: Syst. Control Lett. **14**(2), 145 (1990)
3. Edwards, C., Spurgeon, S.: Sliding Mode Control: Theory and Applications. Series in Systems and Control (Taylor & Francis, Milton Park, 1998)
4. Janardhanan, S., Bandyopadhyay, B.: IEEE Trans. Autom. Control **51**(6), 1030 (2006)
5. Emel'yanov, S.: IEEE Trans. Autom. Control 983–991 (1959)
6. Young, K.: Variable Structure Control for Robotics and Aerospace Applications. Studies in automation and control. Elsevier, Amsterdam (1993)
7. Bandyopadhyay, B., Janardhanan, S.: Discrete-time Sliding Mode Control: A Multirate Output Feedback Approach. Lecture Notes in Control and Information Sciences. Springer, Berlin (2005)
8. Utkin, V.: IEEE Trans. Autom. Control **22**(2), 212 (1977)
9. Arscott, F., Filippov, A.: Differential Equations with Discontinuous Righthand Sides: Control Systems. Mathematics and its Applications. Springer, Netherlands (1988)
10. Gao, W., Wang, Y., Homaifa, A.: IEEE Trans. Ind. Electron. **42**(2), 117 (1995)
11. Bartoszewicz, A.: In: 2017 18th International Carpathian Control Conference (ICCC), pp. 588–593 (2017). https://doi.org/10.1109/CarpathianCC.2017.7970468
12. Milosavljević, C.: Autom. Remote. Control **46**, 307 (1985)
13. Dote, Y., Hoft, R.: Presented at the IAS annual Meeting (1980)
14. Sarpturk, S., Istefanopulos, Y., Kaynak, O.: IEEE Trans. Autom. Control **32**(10), 930 (1987)
15. Hui, S., Żak, S.H.: Syst. Control Lett. **38**(4), 283 (1999)
16. Pai, M.C.: Proc. Inst. Mech. Eng. Part I J. Syst. **226**(7), 927 (2012). https://doi.org/10.1177/0959651812445248
17. Papoulis, A., Pillai, S.U.: Probability, Random Variables, and Stochastic Processes, 4th edn. McGraw Hill, Boston (2002)
18. Zheng, F., Cheng, M., Gao, W.B.: Syst. Control Lett. **22**(3), 209 (1994)
19. Chung, K.L.: A Course in Probability Theory, 2nd edn. No. 21 in Probability and Mathematical Statistics. Academic Press, New York (1974)
20. Kennedy, W.J., Gentle, J.E.: Statistical Computing. Marcel Dekker, New York (1980)
21. Fortmann, T., Williamson, D.: IEEE Trans. Autom. Control **17**(3), 301 (1972)
22. Luenberger, D.: IEEE Trans. Autom. Control **11**(2), 190 (1966)
23. Munro, N.: Proc. Inst. Electr. Eng. **120**(2), 319 (1973)
24. Satyanarayana, N., Janardhanan, S.: Asian J. Control **16**(6), 1897 (2014)
25. Postlethwaite, S.S.I.: Multivariable Feedback Control: Analysis and Design. Wiley India Pvt Ltd, New York (2014)
26. Kalman, R.E., Bucy, R.S.: Trans. ASME Ser. D J. Basic Eng. 109 (1961)
27. Nagpal, K.M., Helmick, R.E., Sims, C.S.: Int. J. Control **45**(6), 1867 (1987)
28. Sage, A., Melsa, J.: Estimation Theory with Applications to Communications and Control. McGraw-Hill series in systems science. McGraw-Hill, New York (1971)
29. Lin, H., Su, H., Shu, Z., Wu, Z.G., Xu, Y.: IEEE Trans. Autom. Control **61**(7), 1794 (2016). https://doi.org/10.1109/TAC.2015.2479195
30. Lu, R., Xu, Y., Zhang, R.: IEEE Trans. Ind. Electron. **63**(11), 6999 (2016). https://doi.org/10.1109/TIE.2016.2585543
31. Xu, Y., Lu, R., Shi, P., Li, H., Xie, S.: IEEE Trans. Cybern. **PP**(99), 1 (2017). https://doi.org/10.1109/TCYB.2016.2635122
32. Song, H., Chen, S.C., Yam, Y.: IEEE Trans. Cybern. **PP**(99), 1 (2017). https://doi.org/10.1109/TCYB.2016.2577340
33. Xu, Y., Lu, R., Peng, H., Xie, K., Xue, A.: IEEE Trans. Neural Netw. Learn. Syst. **28**(2), 268 (2017)
34. Niu, Y., Ho, D.W.C.: IEEE Trans. Autom. Control **55**(11), 2623 (2010). https://doi.org/10.1109/TAC.2010.2069350

35. Argha, A., Li, L., Su, S., Nguyen, H.: IET Control Theory Appl. **10**(11), 1269 (2016). https://doi.org/10.1049/iet-cta.2015.0859
36. Argha, A., Su, S., Li, L., Nguyen, H., Celler, B.: Advances in Discrete-Time Sliding Mode Control: Theory and Applications. CRC Press, Boca Raton. https://books.google.co.in/books?id=0llgDwAAQBAJ (2018)
37. Mehta, A.J., Bandyopadhyay, B.: J. Dyn. Syst. Meas. Contro ASME **138**, 124503 (2016)
38. Kondo, E., Takata, M.: Bull. JSME **20**(142), 428 (1977). https://doi.org/10.1299/jsme1958.20.428
39. Darouach, M.: IEEE Trans. Autom. Control **45**(5), 940 (2000)
40. Trinh, H., Fernando, T.: Functional Observers for Dynamical Systems, vol. 420. Springer, Berlin (2012)
41. Trinh, H., Fernando, T., Emami, K., Huong, D.C.: In: 2013 IEEE 8th International Conference on Industrial and Information Systems, pp. 197–200 (2013)
42. Frank, P.M.: Automatica **26**(3), 459 (1990)
43. Nazmi, S., Mohajerpoor, R., Abdi, H., Nahavandi, S.: In: 2015 IEEE Conference on Control Applications (CCA), pp. 620–625 (2015). https://doi.org/10.1109/CCA.2015.7320698
44. Hou, M., Muller, P.C.: Int. J. Control **60**(5), 827 (1994)
45. Park, T.G.: IET Control Theory Appl. **4**(12), 2781 (2010)
46. Ha, Q., Trinh, H.: Int. J. Syst. Sci. **35**(12), 719 (2004)
47. Xiong, Y., Saif, M.: Automatica **39**(8), 1389 (2003)
48. Ha, Q., Trinh, H., Nguyen, H., Tuan, H.: IEEE Trans. Ind. Electron. **50**(5), 1030 (2003)
49. Aldeen, M., Trinh, H.: IEE Proc. Control Theory Appl. **146**(5), 399 (1999)
50. Rotella, F., Zambettakis, I.: Automatica **47**(1), 164 (2011)

Chapter 2
Design of Sliding Mode Control for Discrete-Time Stochastic Systems with Bounded Disturbances

Abstract In this chapter, the problem of stochastic sliding mode control (SSMC) systems with bounded disturbances is considered and the corresponding SSMC strategies with complete and incomplete state information in the presence of bounded disturbances is established. Using the Kalman filter, we estimate incomplete state information. Also, a stability analysis of bounded disturbances is established. A simulation example is provided to demonstrate the usefulness of the proposed design method.

Keywords Stochastic sliding mode control · Bounded disturbances · Stochastic sliding mode · Discrete time systems · Kalman filter

2.1 Introduction

As mentioned in Chap. 1, the notion of sliding mode control (SMC) for discrete-time stochastic systems has been researched in detail [1]. However, in [1], no synthesis procedure is provided to DSMC for stochastic systems with bounded disturbances. The first contribution of this chapter is the problem of SMC for stochastic systems with bounded disturbances, which is considered along with stability analysis. However, for most practical applications, the full state information is not always available, and only the output is accessible for measurement. In this situation, the Kalman filter has solved the estimation problem in noisy linear systems as a linear filter by use of minimum mean square error [2, 3]. The second contribution of this chapter is the extension of the corresponding SMC policy with complete state information to systems with incomplete state information in the presence of bounded disturbances.

The algorithms proposed in this chapter attempt to derive SMC for discrete-time stochastic systems with bounded disturbances for the complete and incomplete state information cases.

Subsequent sections are organised as follows. Section 2.2 presents the problem formulation after this introductory section. In Sect. 2.3, SMC for systems with complete and incomplete state information synthesis are provided in the presence of bounded disturbances. Stability analysis is given in Sect. 2.4. Detailed numerical

© Springer Nature Switzerland AG 2020

S. Singh and S. Janardhanan, *Discrete-Time Stochastic Sliding Mode Control Using Functional Observation*, Lecture Notes in Control and Information Sciences 483,
https://doi.org/10.1007/978-3-030-32800-9_2

simulation results are presented in Sect. 2.5. Finally, Sect. 2.6 summarises the contributions made in this chapter.

2.2 Problem Formulation

Consider the discrete-time LTI stochastic system

$$x(k + 1) = Ax(k) + Bu(k) + \tilde{d}(k) + \Gamma w(k), \tag{2.1}$$

where $x(k) \in \mathbb{R}^n$ and $u(k) \in \mathbb{R}^m$; $w(k) \in \mathbb{R}^{\tilde{r}}$ is process noise of zero mean with covariance matrix Q. The matrices $A \in \mathbb{R}^{n \times n}$, $B \in \mathbb{R}^{n \times m}$ and $\Gamma \in \mathbb{R}^{n \times \tilde{r}}$ are known constant matrices; $\tilde{d}(k) \in \mathbb{R}^n$ is an unknown but bounded external disturbances of zero mean with covariance matrix V_d.

Assumption 2.1 An uncertain system of the above form is said to have a matched uncertainty if the condition

$$\tilde{d}(k) \in \rho(B)$$

is satisfied. Here $\tilde{d}(k)$ represents the unmodelled dynamics and the known of external disturbances on the system.

The sliding function is the same as that defined in (1.9).

Assumption 2.2 cB is a nonsingular matrix.

Consider the system described by (2.1). We define a new variable $d_m(k)$ as

$$d_m(k) = c\tilde{d}(k).$$

Since the bounded disturbance is known, we have

$$d_l \leq d_m(k) \leq d_u,$$

where the lower bound d_l and the upper bound d_u are known constants. Furthermore, we introduce the following notation:

$$d_0 = \frac{d_l + d_u}{2},$$

where the average value of $d_m(k)$ is denoted by d_0.

2.3 SMC for Stochastic Systems in the Presence of Bounded Disturbances

The concept of SMC for stochastic systems was proposed in [1], with a control algorithm for both complete and incomplete state information.

2.3.1 SMC for Stochastic Systems with Complete State Information in the Presence of Bounded Disturbances

Here our objective is to determine the control that achieves the same goal as that mentioned in (1.18) for system (2.1).

DSMC is obtained as

$$u(k) = -(cB)^{-1}(cAx(k) + d_0),\qquad(2.2)$$

and the variance is $\sigma_c^2 = c(V_d + \Gamma Q\Gamma^T)c^T$.

Remark 2.1 When the SMC (2.2) is used, the sliding variable $s(k)$ will have the dynamics

$$s(k + 1) = d_m(k) - d_0 + c\Gamma w(k),\qquad(2.3)$$

so the system motion (2.1) is called the stochastic sliding mode (SSM).

2.3.2 SMC for Stochastic Systems with Incomplete State Information in the Presence of Bounded Disturbances

As defined in (1.47), the measurement equation is

$$y(k) = Cx(k) + Gv(k).\qquad(2.4)$$

In the case of incomplete state information, the sliding function is the same as that defined in (1.48).

Assumption 2.3 The pair (A, B) is controllable, and the pair (A, C) is observable.

2.3.2.1 An Algorithm for State Estimation in the Presence of Bounded Disturbances

1. $\hat{x}(0) = x_0$ is initially assumed.
2. Given the noise covariance $P(k - 1)$ of the state vector $x(k - 1)$, the error covariance $P(k)$ is determined as

$$(a)\ P(k|k-1) = AP(k-1)A^T + \Gamma Q\Gamma^T + V_d,$$
$$(b)\ P(k) = [P^{-1}(k|k-1) + C^T(GRG^T)^{-1}C]^{-1}, \qquad (2.5)$$
$$(c)\ \Upsilon(k) = P(k)C^T(GRG^T)^{-1},$$

where $\Upsilon(k)$ is the Kalman gain.

3. Using $\Upsilon(k)$, the state vector of the estimated state $\hat{x}(k)$ can be determined as

$$(a)\ \hat{x}(k|k-1) = A\hat{x}(k-1) + Bu(k-1),$$
$$(b)\ \hat{x}(k) = \hat{x}(k|k-1) + \Upsilon(k)[y(k) - C\hat{x}(k|k-1)]. \qquad (2.6)$$

Remark 2.2 The Kalman gain $\Upsilon(k)$ will decrease if the readings (measurements) match the predicted system state. If the measured values say otherwise, the elements of the matrix $\Upsilon(k)$ become larger.

2.3.2.2 Synthesis of SMC

The estimation error is described by $\tilde{x}(k) \triangleq x(k) - \hat{x}(k)$.

On considering

$$\tilde{m}(k+1) \triangleq cA\hat{x}(k) + cBu(k) + d_0 \qquad (2.7)$$

and

$$\tilde{g}(k+1) \triangleq c\Upsilon(k+1)[CA\tilde{x}(k) + C\tilde{d}(k) + C\Gamma w(k) + Gv(k+1)] - d_0, \quad (2.8)$$

applying (2.6) directly to (1.48) yields

$$\hat{s}(k+1) = \tilde{m}(k+1) + \tilde{g}(k+1). \qquad (2.9)$$

By the above algorithm, $\tilde{g}(k+1)$ is Gaussian with mean zero and variance

$$\tilde{\sigma}_c^{\,2}(k+1) = c\Upsilon(k+1)[C(AP(k)A^T + \Gamma Q\Gamma^T + V_d)C^T + GRG^T]\Upsilon^T(k+1)c^T. \qquad (2.10)$$

The DSMC is obtained as

$$u(k) = -(cB)^{-1}(cA\hat{x}(k) + d_0). \qquad (2.11)$$

Remark 2.3 When SMC (2.11) is used, the sliding variable $s(k)$ will have the dynamics

$$\hat{s}(k+1) = c\Upsilon(k+1)[CA\tilde{x}(k) + C\tilde{d}(k) + C\Gamma w(k) + Gv(k+1)] - d_0. \quad (2.12)$$

From (2.9) and (2.12) it can be seen that the conditional variance of $\hat{s}(k+1)$ will be at a minimum when the system reaches a stochastic sliding mode.

2.4 Stability Analysis

2.4.1 Complete State Information Case

The dynamics equation can be represented as

$$x(k+1) = Ax(k) + Bu(k) + \tilde{d}(k) + \Gamma w(k),$$
$$x(k+1) = (A + BL)x(k) - (cB)^{-1}Bd_0 + \tilde{d}(k) + \Gamma w(k),$$

(2.13)

where $L = -(cB)^{-1}cA$.

2.4.2 Incomplete State Information Case

The dynamics equation of the Kalman filter estimate $\hat{x}(k)$ and the estimation error $\tilde{x}(k)$ can be represented as

$$\hat{x}(k+1) = (A + BL)\hat{x}(k) + \Upsilon C\tilde{x}(k) + \Upsilon Gv(k),$$
$$\tilde{x}(k+1) = (A - \Upsilon C)\tilde{x}(k) + \Gamma w(k) + \tilde{d}(k) - \Upsilon Gv(k),$$

(2.14)

where $\Upsilon(k)$ is the Kalman gain obtained from (2.5), which can be given as

$$\Upsilon(k) = [P_0^{-1} + C^T(GRG^T)^{-1}C]^{-1}C^T(GRG^T)^{-1},$$

where P_0 is the unique positive definite solution of the discrete-time algebraic Riccati equation (DARE)

$$P_0 = AP_0A^T - A^T P_0 B[GRG^T + B^T P_0 B]^{-1}B^T P_0 A + \Gamma Q\Gamma^T.$$

2.5 Simulation Results

In this section, we provide an example to illustrate the validity of the proposed approach. The parameters of the matrices of system (2.1) are

$$A = \begin{bmatrix} 0.9944 & -0.1203 & -0.4302 \\ 0.0017 & 0.9902 & -0.0747 \\ 0 & 0.8187 & 0 \end{bmatrix}, B = \Gamma = \begin{bmatrix} 0.4252 \\ -0.0082 \\ 0.1813 \end{bmatrix}, G = \begin{bmatrix} 1 \\ 1 \end{bmatrix}, C = \begin{bmatrix} 1 & 0 & 0 \\ 0 & 1 & 0 \end{bmatrix}.$$

The sliding function is chosen as

$$c = [1 \quad 1 \quad 1].$$

Consider $d(k) = \sin(k/2)e^{-k/5}$, a disturbance of the system. The initial conditions are given for the complete and incomplete state cases $[2 \quad 2 \quad 2]$ and $[-1 \quad -1 \quad -1]$, respectively.

The simulation results are shown in Figs. 2.1, 2.2 and 2.3. Among them, Fig. 2.1 shows the time responses of different state trajectories in the presence of a bounded disturbance. Figure 2.2 shows the control response and sliding function for complete state information in the presence of a bounded disturbance. Figure 2.3 shows the control input and sliding function for incomplete state information in the presence of a bounded disturbance. It is seen that the state trajectory converges to the sliding mode band. These simulation results demonstrate that our proposed design is very effective (Table 2.1).

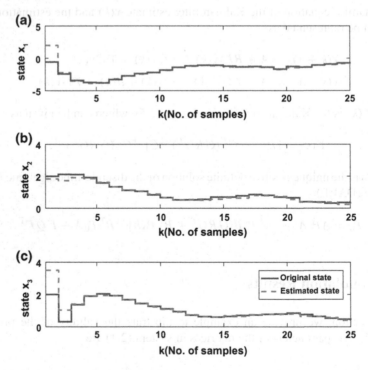

Fig. 2.1 Complete state information in the presence of bounded disturbances: **a–c** time response of the states

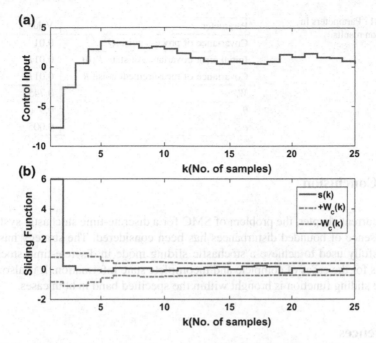

Fig. 2.2 Complete state information in the presence of bounded disturbances: **a** Control input and **b** sliding function

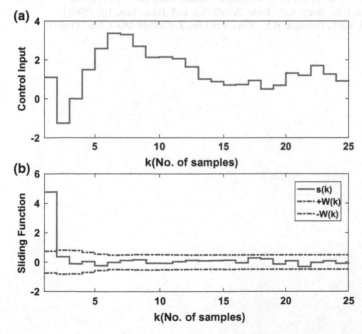

Fig. 2.3 Incomplete state information in the presence of bounded disturbances: **a** Control input and **b** sliding function

Table 2.1 Parameters in simulation results

Parameter	Value
Covariance of process noise Q	0.01
Initial error covariance of state $P(0)$	$0.01 I_3$
Covariance of measurement noise R	$0.01 I_3$
W_c	0.745
θ	0.9
ε	0.005

2.6 Conclusion

In the current chapter, the problem of SMC for a discrete-time stochastic system in the presence of bounded disturbances has been considered. The strategy has been successfully used to achieve a stochastic sliding mode in discrete-time stochastic systems for the cases of complete and incomplete state information. It is also noted that the sliding function is brought within the specified band in both cases.

References

1. Zheng, F., Cheng, M., Gao, W.B.: Syst. Control Lett. **22**(3), 209 (1994)
2. Kalman, R.E., Bucy, R.S.: Trans. ASME Ser. D J. Basic Eng, 109 (1961)
3. Nagpal, K.M., Helmick, R.E., Sims, C.S.: Int. J. Control **45**(6), 1867 (1987)

Chapter 3
Functional Observer-Based Sliding Mode Control for Discrete-Time Stochastic Systems

Abstract This chapter presents a new approach to designing a functional observer-based sliding mode control for discrete-time stochastic systems. The functional observer is based on the Kronecker product approach. Existence conditions and stability analysis of the proposed observer are given. The control input is estimated by a novel linear functional observer. This approach leads to a nonswitching type of control, thereby eliminating the fundamental cause of chatter. Furthermore, the functional observer is designed in such a way that the effect of the process and measurement noise is minimised. A simulation example is given to illustrate and validate the proposed design method.

Keywords Discrete-time systems · Stochastic systems · Sliding mode control · Linear functional observers · State estimation

3.1 Introduction

Designing the simplest possible order observer to estimate a given linear function of a state vector is beneficial from the design point of view [1]. Linear functional observers estimate the linear functions of the state vector [2]. This salient feature has motivated researchers around the world to find ways to design minimum-order functional observers systematically. There exists a method of functional observer-based SMC for an uncertain dynamical system [3], but the problem of the functional observer-based SMC for discrete-time stochastic systems has not been investigated so far. Therefore, it is crucial to extend the functional observer-based SMC theory to stochastic systems.

Motivated by the preceding discussion, this chapter explores the problem of designing functional observer-based SMC for discrete-time stochastic systems.

© Springer Nature Switzerland AG 2020   33
S. Singh and S. Janardhanan, *Discrete-Time Stochastic Sliding Mode Control Using Functional Observation*, Lecture Notes in Control and Information Sciences 483,
https://doi.org/10.1007/978-3-030-32800-9_3

The novelty of the chapter comes from two aspects:

1. A functional observer-based SMC method is proposed for linear discrete-time stochastic systems based on the Kronecker product approach. Existence conditions and stability analysis of functional observers are explored.
2. The SMC is calculated by the functional observer method. This approach leads to a nonswitching-type control.

The chapter is organised as follows. Section 3.2 presents the problem formulation. The sliding function and controller design method are given in Sect. 3.3. The functional observer-based SMC is presented in Sect. 3.4. Detailed numerical simulation results are presented in Sect. 3.5. Finally, Sect. 3.6 summarises the contributions made in this chapter.

3.2 Problem Formulation

A discrete-time LTI stochastic system is described as follows [4]:

$$x(k + 1) = Ax(k) + Bu(k) + \Gamma w(k),$$
$$y(k) = Cx(k) + Gv(k), \tag{3.1}$$

where $x(k) \in \mathbb{R}^n$, $u(k) \in \mathbb{R}^m$ and $y(k) \in \mathbb{R}^p$. The matrices $A \in \mathbb{R}^{n \times n}$, $B \in \mathbb{R}^{n \times m}$, $C \in \mathbb{R}^{p \times n}$, $G \in \mathbb{R}^{p \times l}$, and $\Gamma \in \mathbb{R}^{n \times \bar{r}}$ are known constant matrices, $w(k) \in \mathbb{R}^{\bar{r}}$ is process noise, and $v(k) \in \mathbb{R}^l$ is measurement noise. Let the initial state x_0 be a random vector with zero mean and given covariance matrix P_0. The covariance of plant noise is Q and covariance of measurement noise is R.

Assumption 3.1 The initial state $x(0)$ is a Gaussian random vector, and $x(0)$, $w(k)$, and $v(k)$ are mutually uncorrelated.

Assumption 3.2 The system (3.1) is controllable and observable.

The main task is to design a functional observer-based SMC law for a system (3.1) such that the sliding mode is achieved and the sliding function will lie within a specified band.

3.3 Sliding Function and Controller Design

3.3.1 Design of Sliding Function

The sliding function is the similar to that defined in (1.9) for the system (3.1).

3.3.1.1 Stability Analysis

To obtain the regular form of (3.1), an orthonormal matrix $U \in \mathbb{R}^{n \times n}$ can be selected such that

$$UAU^T = \begin{bmatrix} \bar{A}_{11} & \bar{A}_{12} \\ \bar{A}_{21} & \bar{A}_{22} \end{bmatrix}, UB = \begin{bmatrix} 0_{(n-m) \times m} \\ B_2 \end{bmatrix}, U\Gamma = \begin{bmatrix} 0 \\ \Gamma_2 \end{bmatrix},$$

where $B_2 \in \mathbb{R}^{m \times m}$ is nonsingular. By the transformation of state $\xi(k) = Ux(k)$, (3.1) has the regular form

$$
\begin{aligned}
\xi_1(k+1) &= \bar{A}_{11}\xi_1(k) + \bar{A}_{12}\xi_2(k), \\
\xi_2(k+1) &= \bar{A}_{21}\xi_1(k) + \bar{A}_{22}\xi_2(k) + B_2 u(k) + \Gamma_2 w(k),
\end{aligned}
\tag{3.2}
$$

where $\xi_1(k) \in \mathbb{R}^{n-m}$ and $\xi_2(k) \in \mathbb{R}^m$.

The sliding function (1.9) can be expressed in terms of the new state $\xi(k)$ as

$$s(k) \triangleq cU^T \xi(k) \triangleq [K \quad I_m]\xi(k) = K\xi_1(k) + \xi_2(k). \tag{3.3}$$

Substituting $\xi_2(k) = -K\xi_1(k)$ in the first equation of system (3.2) gives the dynamics

$$\xi_1(k+1) = (\bar{A}_{11} - \bar{A}_{12}K)\xi_1(k), \tag{3.4}$$

where the matrix K is chosen such that $(\bar{A}_{11} - \bar{A}_{12}K)$ is stable.

The sliding surface (1.9) can be expressed in terms of the original state coordinates as

$$s(k) = cx(k) = cU^T \xi(k) = 0. \tag{3.5}$$

3.3.2 Synthesis of Control Law

Our objective is find the control that achieves the same objective as that in (1.18) for the system (3.1). The analysis of the controller design is similar to that in Chap. 1.

The control can be obtained as

$$u(k) = -(cB)^{-1}cAx(k). \tag{3.6}$$

The stochastic system state (3.1) will lie in $S_{\mu c}$, which can be shown using (3.6) and $\sigma_c^2 = c\Gamma Q\Gamma^T c^T$.

3.4 Linear Functional Observer-Based SMC Design

In this section, we design an observer for estimating the linear functional of the states. Let $u(k)$ be a vector that is required to be estimated, where

$$u(k) = Lx(k), \tag{3.7}$$

and $L \in \mathbb{R}^{m \times n}$ is a known matrix. The matrix L, which represents any desired partial set of the state vector to be estimated, can always be obtained by the SMC (3.6).

The aim is to design an observer of the form

$$\zeta(k+1) = M\zeta(k) + Jy(k) + Hu(k),$$
$$\hat{u}_f(k) = V\zeta(k) + Ey(k), \tag{3.8}$$

where $\zeta(k) \in \mathbb{R}^q$ is the observer state vector; $\hat{u}_f(k)$ is the estimate of the $u(k)$ functional state vector, and $M \in \mathbb{R}^{q \times q}$, $J \in \mathbb{R}^{q \times p}$, $H \in \mathbb{R}^{q \times m}$, $V \in \mathbb{R}^{r \times q}$, and $E \in \mathbb{R}^{r \times p}$ are unknown constant matrices.

Remark 3.1 The observer dynamics are determined by the order of the matrix M, which may be chosen arbitrarily, with the restriction that its eigenvalues must be distinct, different from the eigenvalues of A, and lie inside the unit disk.

Theorem 3.1 *The qth-order observer (3.8) will estimate $u_f(k)$ if the following conditions hold:*

  (i) M is a stable matrix;
 (ii) $VT = L - EC$;
(iii) $MT + JC - TA = 0$;
 (iv) $H = TB$;
  (v) $q \geq \rho(L(I_n - C^+C))$,

where $L \in \mathbb{R}^{r \times n}$ is the functional gain matrix and $T \in \mathbb{R}^{q \times n}$ is the unknown constant matrix.

Proof Let us define $e(k)$, the error between $\zeta(k)$ and $Tx(k)$, as

$$e(k) \triangleq \zeta(k) - Tx(k). \tag{3.9}$$

After taking the first-order difference of (3.9), this gives

$$
\begin{aligned}
e(k+1) &= \zeta(k+1) - Tx(k+1) \\
&= M\zeta(k) + Jy(k) + Hu(k) - TAx(k) - TBu(k) - T\Gamma w(k) \\
&= Me(k) + (MT + JC - TA)x(k) + (H - TB)u(k) + JGv(k) - T\Gamma w(k).
\end{aligned}
\tag{3.10}
$$

The independence of the error dynamics (3.10) from $x(k)$, $u(k)$ requires that

$$MT + JC - TA = 0 \tag{3.11}$$

and

$$H - TB = 0. \tag{3.12}$$

Subject to (3.11) and (3.12), the dynamics of the error (3.10) becomes

$$e(k+1) = Me(k) + JGv(k) - T\Gamma w(k). \tag{3.13}$$

Hence the error dynamics $e(k)$ is governed by the matrices M, J, and T.
   The output estimation error is expressed as

$$
\begin{aligned}
e_u(k) &\triangleq \hat{u}_f(k) - Lx(k) \\
&= V\zeta(k) + Ey(k) - Lx(k) \\
&= Ve(k) + (VT + EC - L)x(k) + EGv(k).
\end{aligned}
\tag{3.14}
$$

The independence of the output error dynamics (3.14) from $x(k)$ requires that

$$VT = L - EC. \tag{3.15}$$

Then the output estimation error becomes

$$e_u(k) = Ve(k) + EGv(k). \tag{3.16}$$

Hence the output error dynamics (3.16) is governed by matrices V and E. The matrix E can be chosen as $E = LC^+$ to satisfy the observer condition (v) of Theorem 3.1.

Remark 3.2 In the given system (3.1), if $w(k)$ and $v(k)$ are zero, then the discrete-time stochastic system acts as a discrete-time system. In that case, if conditions (iii) and (iv) of Theorem 3.1 are satisfied, then (3.13) is reduced to $e(k+1) = Me(k)$. Further, if condition (ii) is satisfied, then $e_u(k)$ is reduced to $Ve(k)$. Since $e(k) \to 0$ as $k \to \infty$, it follows that $e_u(k) \to 0$, and hence $\hat{u}(k) \to u(k)$ as $k \to \infty$.

Further, the unknown terms J and T in observer Eq. (3.8) can be solved as follows. Let

$$[J\ T] = \mathscr{X}, \tag{3.17}$$

where $\mathscr{X} \in \mathbb{R}^{q \times (n+p)}$ is an unknown matrix.
   Using (3.17), the matrices J and T can be expressed in terms of the unknown matrix $\mathscr{X}$ as

$$J = \mathscr{X} \begin{bmatrix} I_{p \times p} \\ 0_{n \times p} \end{bmatrix} \tag{3.18}$$

and

$$T = \mathscr{X} \begin{bmatrix} 0_{p \times n} \\ I_{n \times n} \end{bmatrix}. \tag{3.19}$$

On substituting for J and T in (3.11),

$$M \mathscr{X} \begin{bmatrix} 0 \\ I \end{bmatrix} + \mathscr{X} \begin{bmatrix} I \\ 0 \end{bmatrix} C - \mathscr{X} \begin{bmatrix} 0 \\ I \end{bmatrix} A = 0$$

or

$$M \mathscr{X} \begin{bmatrix} 0 \\ I \end{bmatrix} + \mathscr{X} \begin{bmatrix} C \\ -A \end{bmatrix} = 0 \tag{3.20}$$

$$V \mathscr{X} \begin{bmatrix} 0 \\ I \end{bmatrix} = L - EC, \tag{3.21}$$

now augmenting the matrices (3.20) and (3.21) in a composite form would give

$$\begin{bmatrix} M \\ V \end{bmatrix} \mathscr{X} \begin{bmatrix} 0 \\ I \end{bmatrix} + \begin{bmatrix} I \\ 0 \end{bmatrix} \mathscr{X} \begin{bmatrix} C \\ -A \end{bmatrix} = \Theta \tag{3.22}$$

with

$$\Theta = \begin{bmatrix} 0 \\ L - EC \end{bmatrix},$$

where $\Theta \in \mathbb{R}^{(q+r) \times n}$ is a known constant matrix.

Applying the Kronecker product [5] in (3.22) gives

$$\Sigma vec(\mathscr{X}) = vec(\Theta) \tag{3.23}$$

with

$$\Sigma = \begin{bmatrix} \begin{bmatrix} 0 \\ I \end{bmatrix}^T \otimes \begin{bmatrix} M \\ V \end{bmatrix} + \begin{bmatrix} C \\ -A \end{bmatrix}^T \otimes \begin{bmatrix} I \\ 0 \end{bmatrix} \end{bmatrix},$$

where $vec(\mathscr{X}) \in \mathbb{R}^{(n+p)q}$, $vec(\Theta) \in \mathbb{R}^{n(q+r)}$, and $\Sigma \in \mathbb{R}^{(q+r)n \times q(n+p)}$.

The error covariance of (3.13) propagates as

$$P(k+1) = MP(k)M^T + JGRG^T J^T + T\Gamma Q\Gamma^T T^T. \tag{3.24}$$

The output error covariance of (3.16) is

$$P_u(k) = VP(k)V^T + EGRG^T E^T. \tag{3.25}$$

Now we can write $P_u(k+1)$ in terms of the covariance as

$$P_u(k+1) = V(MP(k)M^T + JGRG^T J^T + T\Gamma Q\Gamma^T T^T)V^T + EGRG^T E^T.$$
$$(3.26)$$

To minimise the effect of the process and measurement noise covariance, we choose $\mathscr{X}$ in (3.17) as

$$\mathscr{X} = \underset{\mathscr{X}}{\operatorname{argmin}} \left\| V\mathscr{X} \begin{bmatrix} GRG^T & 0 \\ 0 & \Gamma Q\Gamma^T \end{bmatrix} \mathscr{X}^T V^T \right\| \qquad (3.27)$$

to minimise the effect of noise in the functional observer.

If condition (v) is satisfied, then we can design a stable q-dimensional system, where

$$q \geq \rho(L(I_n - C^+C)),$$

to generate any required vector state function of $x(k)$. This proves condition (v) and completes the proof of Theorem 3.1.

Based on the procedure explained above, the design policy is now summarised as follows:

Algorithm 3.1 Functional observer-based SMC for stochastic systems

---

1: To obtain the sliding function $s(k) \triangleq cx(k)$.
2: Design controller $u(k)$, and the functional to be estimated is $Lx(k)$, where $L = -(cB)^{-1}cA$.
3: Choose the $(r \times q)$ elements of the matrix V arbitrarily.
4: Obtain the minimum order of the functional observer $q \geq \rho(L(I_n - C^+C))$ such that $\rho(V) = \rho(L - EC)$.
5: Choose arbitrarily a stable $(q \times q)$ matrix M.
6: Solve the composite form of matrix (3.22) by the Kronecker product for the matrix $\mathscr{X}$. The matrix $\mathscr{X}$ is chosen according to

$$\mathscr{X} = \underset{\mathscr{X}}{\operatorname{argmin}} \left\| \mathscr{X} \begin{bmatrix} GRG^T & 0 \\ 0 & \Gamma Q\Gamma^T \end{bmatrix} \mathscr{X}^T \right\|,$$

subject to the equality condition (3.22). If satisfied, go to the next step; otherwise set $q = q + 1$ and go to step 4.
7: From (3.18) and (3.19), find the values of the matrices J and T.
8: Now find the value of the matrix H by $H = TB$.
9: As a result of steps 1 to 8, obtain the structure of the functional observer as in (3.8).

3.5 Simulation Results

Consider the unstable multi-input–multi-output linear discrete-time stochastic system (3.1), where the system parameter matrices are

$$A = \begin{bmatrix} 0 & 0 & 0 & 1 & 0 & 0 & 1 \\ 1 & 0 & 0 & 0 & 0 & 1 & 0 \\ 0 & 0 & 0 & 0 & 1 & 0 & 0 \\ 0 & 0 & 1 & 0 & 0 & 0 & 1 \\ 0 & 2 & 1 & 1 & 0 & 1 & 0 \\ 0 & 0 & 0 & 0 & 0 & 0 & 1 \\ 0 & 1 & 0 & -1 & 0 & 0 & 0 \end{bmatrix}, B = \begin{bmatrix} 0 & 0 \\ 0 & 1 \\ 1 & 0 \\ 0 & 1 \\ 0 & 0 \\ 1 & 0 \\ 0 & 0 \end{bmatrix}, \Gamma = \begin{bmatrix} 0 \\ 0 \\ 0 \\ 0 \\ 1 \\ 0 \\ 1 \end{bmatrix},$$

$$C = \begin{bmatrix} I_3 & 0_4 \end{bmatrix}, G = \begin{bmatrix} 1 & 1 & 1 \end{bmatrix}^T.$$

The open-loop unstable eigenvalues of the matrix A are obtained as $eig(A) = 0.141, -0.316 \pm 1.522i, -1.058 \pm 0.722i, 1.303 \pm 0.299i$ [6, pp 649–650]. The covariance of the process noise and measurement noises are $Q = 0.1I$ and $R = 0.1I$, respectively [7]. The parameters θ and ε are chosen as $\theta = 0.9$ and $\varepsilon = 0.01$. The initial conditions of the original system and observer are chosen arbitrarily as $x(0) = \begin{bmatrix} 1 & 3 & 6 & 4 & 8 & 5 & 7 \end{bmatrix}^T$ and $\zeta(0) = \begin{bmatrix} 1 & 2 & 7 \end{bmatrix}^T$.

The sliding function is obtained as

$$s(k) = \begin{bmatrix} 0.867 & -0.459 & 0.933 & -1.729 & -0.589 & 0.578 & -0.826 \\ -0.305 & -0.205 & -1.375 & -0.175 & 0.837 & -0.539 & -0.839 \end{bmatrix} x(k).$$

The control input can be shown to be

$$u(k) = \begin{bmatrix} -0.260 & -1.681 & -1.848 & 0.348 & 1.094 & -0.941 & 0.121 \\ -0.129 & 0.529 & 0.418 & 0.866 & -0.869 & 0.399 & -0.645 \end{bmatrix} x(k).$$

Now E can be computed as

$$E = \begin{bmatrix} -0.260 & -1.681 & -1.848 \\ -0.129 & 0.529 & 0.418 \end{bmatrix}.$$

Here the order of the observer is obtained as $q = 3$.

A $(q \times q)$ diagonal convergent matrix is chosen as

$$M = \begin{bmatrix} 0.2 & 0 & 0 \\ 0 & 0.1 & 0 \\ 0 & 0 & 0.3 \end{bmatrix}.$$

Consider an arbitrarily chosen matrix V:

$$V = \begin{bmatrix} 3 & 1 & 2 \\ 4 & 2 & 5 \end{bmatrix}.$$

The matrix $\mathscr{X}$ is now obtained by (3.27) as $\mathscr{X} = \begin{bmatrix} \mathscr{X}_1 & \mathscr{X}_2 \end{bmatrix}$, where

$$\mathscr{X}_1 = \begin{bmatrix} -0.059 & 0.445 & 0.063 & -0.426 & 0.014 \\ -0.306 & -2.067 & 0.678 & 2.231 & 0.017 \\ 0.081 & 0.318 & -0.333 & -0.417 & 0.019 \end{bmatrix}$$

and

$$\mathscr{X}_2 = \begin{bmatrix} 0.440 & 0.029 & -0.011 & 0.187 & -0.126 \\ -1.694 & -0.476 & 0.077 & 0.079 & 0.606 \\ 0.304 & 0.172 & -0.022 & -0.203 & -0.137 \end{bmatrix}.$$

Solving (3.18) for the matrix J gives

$$J = \begin{bmatrix} -0.059 & 0.445 & 0.063 \\ -0.307 & -2.067 & 0.678 \\ 0.081 & 0.318 & -0.333 \end{bmatrix},$$

and solving (3.19) for the matrix T gives

$$T = \begin{bmatrix} 0.440 & 0.029 & -0.011 & 0.187 & -0.126 & -0.759 & 0.703 \\ -1.694 & -0.476 & 0.077 & 0.079 & 0.606 & 1.262 & -3.327 \\ 0.304 & 0.172 & -0.022 & -0.203 & -0.137 & 0.161 & 0.644 \end{bmatrix}.$$

Since $H = TB$, the matrix H is obtained as

$$H = \begin{bmatrix} 0.577 & -0.029 \\ -2.720 & -2.540 \\ 0.507 & 0.977 \end{bmatrix}.$$

Thus the estimate of $u(k)$ is given by the functional observer (3.8).

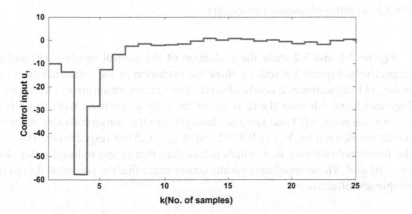

Fig. 3.1 Evolution of control input $u_1(k)$

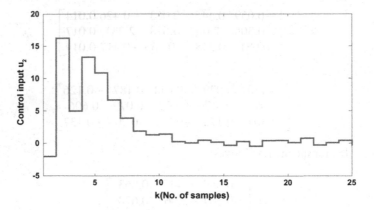

Fig. 3.2 Evolution of control input $u_2(k)$

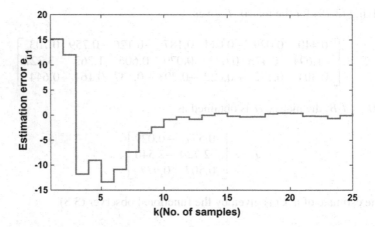

Fig. 3.3 Evolution of estimation error $e_{u1}(k)$

Figures 3.1 and 3.2 show the evaluation of the control inputs $u_1(k)$ and $u_2(k)$, respectively. Figures 3.3 and 3.4 show the evolution of the estimated errors $e_{u1}(k)$ and $e_{u2}(k)$, respectively. It is to be observed that the estimation error converges to zero. Figures 3.5 and 3.6 show the response of the sliding functions $s_1(k)$ and $s_2(k)$. The sliding functions $s_1(k)$ and $s_2(k)$ are brought into the constant sliding mode bands, which are defined by $W_{c1} = 0.8162$ and $W_{c2} = 0.5799$ respectively. The order of the functional observer is 3, which is less than that of the reduced order observer $(n - p) = 4$. These simulation results demonstrate that the proposed design is very simple and effective.

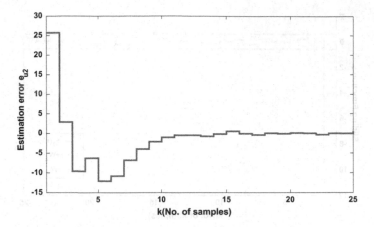

Fig. 3.4 Evolution of estimation error $e_{u2}(k)$

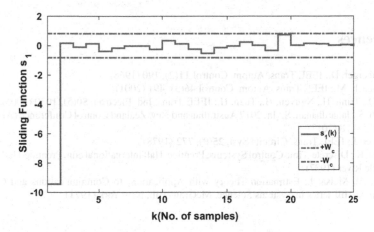

Fig. 3.5 Evolution of sliding function $s_1(k)$

3.6 Conclusion

In this chapter, the problem of functional observer-based SMC for discrete-time stochastic systems has been considered. A strategy has been successfully implemented to find the functional observer in stochastic systems and minimise the effect of noise. Existence conditions and stability analysis of the functional observer have been derived. A numerical example was considered to illustrate the main features of the proposed method.

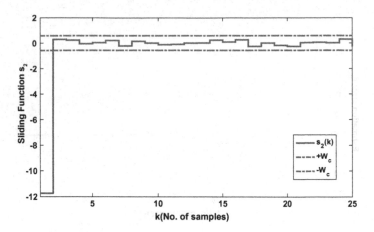

Fig. 3.6 Evolution of sliding function $s_2(k)$

References

1. Luenberger, D.: IEEE Trans. Autom. Control **11**(2), 190 (1966)
2. Darouach, M.: IEEE Trans. Autom. Control **46**(3), 491 (2001)
3. Ha, Q., Trinh, H., Nguyen, H., Tuan, H.: IEEE Trans. Ind. Electron. **50**(5), 1030 (2003)
4. Singh, S., Janardhanan, S.: In: 2017 Australian and New Zealand Control Conference (ANZCC), pp. 175–178 (2017)
5. Brewer, J.: IEEE Trans. Circuits Syst. **25**(9), 772 (1978)
6. Ogata, K.: Discrete-Time Control Systems, Prentice-Hall International edn. Prentice-Hall, Upper Saddle River (1987)
7. Sage, A., Melsa, J.: Estimation Theory with Applications to Communications and Control. McGraw-Hill series in Systems Science. McGraw-Hill, New York (1971)

Chapter 4
Functional Observer-Based Sliding Mode Control for Discrete-Time Stochastic Systems with Unmatched Uncertainty

Abstract A discrete-time sliding mode control is one of the most prominent robust control techniques for stabilising systems. However, the assumption of full state feedback being available is seldom satisfied in practice. This chapter presents the functional observer-based sliding mode controller (SMC) design for discrete-time stochastic systems with unmatched uncertainty. A disturbance-dependent sliding function method is proposed such that the effect of unmatched uncertainty of the system is minimised. Furthermore, SMC is calculated using the functional observer method. Finally, a simulation example is given to show the effectiveness of the proposed method.

Keywords Discrete-time systems · Stochastic systems · Unmatched uncertainty · Sliding mode controller · Linear functional observers

4.1 Introduction

The design of a linear dynamical functional observer to reconstruct a linear function of the state vector for systems with unknown inputs is reported in [1–3]. In most realistic systems, various uncertainties, such as parameter changes and unknown external excitations, can be incorporated as unknown inputs in the system model [4, 5]. The standard Luenberger observer theory [6, 7] has been effectively applied in the design of functional observers for systems with unknown inputs [1]. The presence of unmatched uncertainty may not satisfy the matching condition, wherein these uncertainties affect the system through the control signal [8]. Therefore, these uncertainties cannot be eliminated through the DSMC technique [9, 10]. Thus, significant efforts have been made to design control strategies to counteract the mismatched disturbances, and numerous control strategies have been proposed, such as integral SMC and the matrix Riccati approach [11–16].

© Springer Nature Switzerland AG 2020

S. Singh and S. Janardhanan, *Discrete-Time Stochastic Sliding Mode Control Using Functional Observation*, Lecture Notes in Control and Information Sciences 483, https://doi.org/10.1007/978-3-030-32800-9_4

A functional observer-based SMC design for stochastic systems with unmatched uncertainty has not been addressed in the literature. Motivated by the preceding discussion, this chapter explores the problem of the functional observer-based SMC design for discrete-time stochastic systems with unmatched uncertainty. However, a functional observer-based SMC satisfying a certain condition may be designed for a system with unmatched uncertainty such that the sliding surface is free of the influence of the disturbance. The control input consists of the function of the system state and disturbance components [17].

The novelty of this chapter is the following:

1. A disturbance-dependent sliding surface method is proposed for discrete-time stochastic systems in the presence of unmatched uncertainty. The sliding surface is developed to ameliorate the effect of unmatched uncertainty in the given system.
2. The SMC is calculated by a functional observer method.

The rest of the chapter comprises the following: Section 4.2 presents the concept of stochastic SMC after this introductory section. Sliding function and controller design are presented in Sect. 4.3. Linear functional observer-based SMC is proposed with existence conditions and design analysis in Sect. 4.4. The next section, Sect. 4.5, validates the findings of the chapter through numerical simulations. Lastly, Sect. 4.6 summarises the contributions made in this chapter.

4.2 Problem Formulation

Consider the following LTI stochastic system with unmatched uncertainty:

$$
\begin{aligned}
x(k+1) &= Ax(k) + Bu(k) + Fd(k) + \Gamma w(k), \\
y(k) &= Cx(k) + Gv(k),
\end{aligned}
\tag{4.1}
$$

where $x(k) \in \mathbb{R}^n$, $u(k) \in \mathbb{R}^m$, $d(k) \in \mathbb{R}^{r_d}$, and $y(k) \in \mathbb{R}^p$. It is assumed that $x(0)$ is a Gaussian random vector and $x(0)$, $d(k)$, $w(k)$, and $v(k)$ are mutually uncorrelated. The following assumptions are made for the purpose of the design of the stochastic SMC.

Assumption 4.1 The pairs (A, B) and (A, C) are controllable and observable, respectively.

Assumption 4.2 The matching condition $\rho(B) = \rho(B|\Gamma)$ is satisfied, but the unmatched condition $\rho(B) = \rho(B|F)$ is not necessarily satisfied (unmatched means $\rho(B) < \rho(B|F)$).

The main goal is to design a functional observer-based SMC for the system (4.1) such that the sliding mode is obtained and mismatched uncertainty in the system is minimised.

4.3 Sliding Function and Controller Design

The design problem comprises the following three stages. In the first stage, a sliding function is designed. In the second stage, the appropriate controller inputs are developed, and in the final stage, the matched and unmatched uncertainty is compensated by the appropriate selection of the augmented system.

4.3.1 Sliding Function Design

In this section, we would like to design a sliding function in order to develop an SMC of the system (4.1) such that the sliding motion is stable. Assume that $U \in \mathbb{R}^{n \times n}$ is an orthonormal matrix, so that $UB = [0_{(n-m) \times m} \quad B_2^T]^T$, where B_2 is a nonsingular matrix. Let $\xi(k) = [\xi_1^T(k) \quad \xi_2^T(k)]^T = Ux(k)$. Then the system (4.1) can be brought into regular form

$$
\begin{aligned}
\xi_1(k+1) &= \bar{A}_{11}\xi_1(k) + \bar{A}_{12}\xi_2(k) + F_1 d_{un}(k), \\
\xi_2(k+1) &= \bar{A}_{21}\xi_1(k) + \bar{A}_{22}\xi_2(k) + B_2 u(k) + F_2 d_m(k) + \Gamma_2 w(k),
\end{aligned}
\tag{4.2}
$$

where $\xi_1(k) \in \mathbb{R}^{n-m}$ and $\xi_2(k) \in \mathbb{R}^m$ are blocks of the state vector. The matrices $\bar{A}_{11} \in \mathbb{R}^{(n-m) \times (n-m)}$, $\bar{A}_{12} \in \mathbb{R}^{(n-m) \times m}$, $\bar{A}_{21} \in \mathbb{R}^{m \times (n-m)}$, and $\bar{A}_{22} \in \mathbb{R}^{m \times m}$ are blocks of the system matrix $\bar{A}$; $d_{un}(k)$ and $d_m(k)$ are the unmatched and matched parts of the uncertainty; $F_1 \in \mathbb{R}^{(n-m) \times r_d}$, $F_2 \in \mathbb{R}^{m \times r_d}$ and $\Gamma_2 \in \mathbb{R}^{m \times \bar{r}}$ are blocks of the uncertainty and noise matrix of the system.

Remark 4.1 The system (4.2) in the sliding mode is independent of $d(k)$ if and only if $F_1 = 0$. The system in the sliding mode is free of $d(k)$ if there exists a matrix Ξ such that $F = B\Xi$. If $m = 1$, this condition is also necessary.

The proposed sliding function is expressed as [9]

$$
s(k) \triangleq cx(k) + k_0 F_1 d_{un}(k),
\tag{4.3}
$$

where $s(k) \in \mathbb{R}^m$ is the sliding function, $c \in \mathbb{R}^{m \times n}$ is the gain matrix, and $k_0 \in \mathbb{R}^{m \times (n-m)}$ is a constant matrix.

Remark 4.2 The conventional sliding surface is described as $s(k) = cx(k)$, but in this chapter, the sliding surface is constructed as $s(k) = cx(k) + k_0 F_1 d_{un}(k)$. The proposed sliding surface is dependent on the state and uncertainty, whereas in the conventional sliding surface, only states are involved. Using the proposed sliding surface (4.3), it can minimise the effect of unmatched uncertainty that is presented in the system.

Assumption 4.3 cB is a nonsingular matrix.

The sliding function (4.3) may be rewritten using the state transformation

$$s(k) = K\xi_1(k) + \xi_2(k) + k_0 F_1 d_{un}(k).\tag{4.4}$$

While the system is on the sliding surface (4.4), it can be rewritten as

$$\xi_2(k) = -(K\xi_1(k) + k_0 F_1 d_{un}(k)).$$

Now put the values of $\xi_2(k)$ in (4.2):

$$\xi_1(k+1) = (\bar{A}_{11} - \bar{A}_{12}K)\xi_1(k) + (I_{(n-m)} - \bar{A}_{12}k_0)F_1 d_{un}(k),\tag{4.5}$$

where K is a parameter to be designated later.

Remark 4.3 It is possible to minimise the unmatched uncertainty effect in the sliding mode if $k_0 = \bar{A}_{12}^+$, where $\bar{A}_{12}^+$ is the left pseudoinverse of $\bar{A}_{12}$ [13].

4.3.1.1 Stability Analysis of Sliding Function

Consider the quadratic cost function for the discrete-time system [18]

$$\mathfrak{J} = \sum_{k=0}^{\infty} \xi_1^T(k)\mathfrak{Q}_{11}\xi_1(k) + 2\xi_1^T(k)\mathfrak{Q}_{12}\xi_2(k) + \xi_2^T(k)\mathfrak{Q}_{22}\xi_2(k),\tag{4.6}$$

where

$$\begin{bmatrix}\mathfrak{Q}_{11} & \mathfrak{Q}_{12}\\ \mathfrak{Q}_{21} & \mathfrak{Q}_{22}\end{bmatrix} \geq 0, \quad \mathfrak{Q}_{22} > 0.$$

Then (4.6) is converted to the quadratic optimal regulator

$$\mathfrak{J} = \sum_{k=0}^{\infty} \xi_1^T(k)\bar{\mathfrak{Q}}\xi_1(k) + \eta^T(k)\mathfrak{Q}_{22}\eta(k),\tag{4.7}$$

where $\bar{\mathfrak{Q}} = \mathfrak{Q}_{11} - \mathfrak{Q}_{12}\mathfrak{Q}_{22}^{-1}\mathfrak{Q}_{12}$ and $\eta(k) = \xi_2(k) + \mathfrak{Q}_{22}^{-1}\mathfrak{Q}_{12}\xi_1(k)$.

From (4.2), the nominal system with respect to $x_1(k)$ can be acquired as

$$\xi_1(k+1) = \bar{A}_{11}\xi_1(k) + \bar{A}_{12}\xi_2(k).$$

Consequently, we get

$$\xi_1(k+1) = \hat{A}\xi_1(k) + \bar{A}_{12}\eta(k),\tag{4.8}$$

where $\hat{A} = \bar{A}_{11} - \bar{A}_{12}\mathfrak{Q}_{22}^{-1}\mathfrak{Q}_{12}$. Since $\mathfrak{Q}$ is a positive definite symmetric matrix, it is clear that $\bar{\mathfrak{Q}} > 0$. If Assumption 4.1 holds, then the pair $(\bar{A}_{11}, \bar{A}_{12})$ and the pair

$(\hat{A}, \bar{A}_{12})$ are also controllable. Through minimisation of $\mathfrak{J}$ in association with the nominal system in (4.8), we obtain

$$\eta(k) = -\mathfrak{Q}_{22}^{-1} \bar{A}_{12}^T \mathfrak{P} \xi_1(k), \tag{4.9}$$

where $\mathfrak{P}$ is a positive definite symmetric solution of the DARE

$$\hat{A}^T \mathfrak{P} \hat{A} - \mathfrak{P} - \hat{A}^T \mathfrak{P} \bar{A}_{12} (\bar{A}_{12}^T \mathfrak{P} \bar{A}_{11} + \mathfrak{Q}_{22})^{-1} \bar{A}_{12}^T \mathfrak{P} \hat{A} + \tilde{\mathfrak{Q}} = 0. \tag{4.10}$$

Thus on minimising $\mathfrak{J}$ in (4.6) with respect to $\eta(k)$, one can obtain the optimal sliding gain matrix

$$\xi_2(k) = -(\mathfrak{Q}_{22} + \bar{A}_{12}^T \mathfrak{P} \bar{A}_{12})^{-1} \bar{A}_{12}^T \mathfrak{P} \hat{A} \xi_1(k). \tag{4.11}$$

Using the above equation, the sliding gain matrix K is obtained as

$$K = (\mathfrak{Q}_{22} + \bar{A}_{12}^T \mathfrak{P} \bar{A}_{12})^{-1} (\bar{A}_{12}^T \mathfrak{P} \bar{A}_{11} + \mathfrak{Q}_{21}). \tag{4.12}$$

The matrix K guarantees that all the eigenvalues of $(\bar{A}_{11} - \bar{A}_{12}K)$ lie inside the unit circle.

Assumption 4.4 The presence of a disturbance in the system is slowly varying, i.e., it can be said that the difference between two sampling instances, $|d(k+1) - d(k)|$, is insignificant. Hence the disturbance $d(k)$ would be a good estimate of the disturbance $d(k+1)$ [19].

4.3.2 Augmented System

In this section, we introduce the augmented state vector $x_d(k) = [x^T(k) \quad d^T(k)]^T \in \mathbb{R}^{(n+r_d)}$ and form the augmented system by combining (4.1) and Assumption 4.4. We can rewrite (4.1) as

$$\begin{aligned} x_d(k+1) &= \bar{A}x_d(k) + \bar{B}u(k) + \bar{\Gamma}w(k), \\ y(k) &= \bar{C}x_d(k) + Gv(k), \end{aligned} \tag{4.13}$$

where

$$\bar{A} = \begin{bmatrix} A & F \\ 0 & I \end{bmatrix} \in \mathbb{R}^{(n+r_d)\times(n+r_d)}, \quad \bar{B} = \begin{bmatrix} B \\ 0 \end{bmatrix} \in \mathbb{R}^{(n+r_d)\times m}$$

$$\bar{\Gamma} = \begin{bmatrix} \Gamma \\ 0 \end{bmatrix} \in \mathbb{R}^{(n+r_d)\times \bar{r}}, \quad \bar{C} = \begin{bmatrix} C & 0 \end{bmatrix} \in \mathbb{R}^{p\times(n+r_d)}.$$

4.3.3 Controller Design

After taking the first-order difference of the sliding function (4.3), we obtain

$$s(k + 1) = cx(k + 1) + k_0 F_1 d_{un}(k + 1). \tag{4.14}$$

Putting the values from (4.1) in (4.14) and using Assumption 4.4, we can write

$$m(k + 1) = cAx(k) + cBu(k) + cFd(k) + k_0 F_1 d_{un}(k)$$

and

$$g(k + 1) = c\Gamma w(k) + k_0[I \quad 0]F(d(k + 1) - d(k)),$$

where the variance is given by $\sigma_c^2 = c\Gamma Q\Gamma^T c^T$,

$$m(k + 1) = cAx(k) + cBu(k) + cFd(k) + k_0 F_1 d_{un}(k) = 0.$$

The controller is obtained as

$$u(k) = Lx_d(k), \tag{4.15}$$

where $L = [-(cB)^{-1}cA \quad - (cB)^{-1}(c + k_0[I \quad 0])F]$.

Remark 4.4 When the SMC $u(k) = Lx_d(k)$ is used, we will have

$$s(k + 1) = c\Gamma w(k) + k_0[I \quad 0]F(d(k + 1) - d(k)). \tag{4.16}$$

Since the unmatched uncertainty is bounded, the difference in the uncertainty term will also be bounded.

Discussion: The following conclusions can be drawn from Sect. 4.2.

1. As shown in (4.5), on using the proposed sliding function (4.3), the unmatched uncertainty is multiplied by the projection matrix $(I - \bar{A}_{12}k_0)$. This projection matrix can be designed such that the effect of unmatched uncertainties on $\xi_1(k)$ is minimised.
2. The proposed control can be used to handle the unmatched uncertainties of the stochastic system, which is not possible in the conventional discrete-time stochastic sliding mode.

In the next section, we propose a functional observer-based SMC method for augmented discrete-time stochastic systems.

4.4 Linear Functional Observer-Based SMC Design

In this section, our objective is to estimate a linear function of the state vector of the form $u(k) = Lx_d$, where $x_d(k)$ is the state vector of the discrete-time system (4.13). The control gain matrix L is calculated via the control input (4.15).

Our aim is to design a functional observer of the form

$$\zeta(k+1) = M\zeta(k) + Jy(k) + Hu(k),$$
$$\hat{u}_f(k) = V\zeta(k) + Ey(k), \tag{4.17}$$

where $\zeta(k) \in \mathbb{R}^q$ is the state vector, and $\hat{u}_f(k) \in \mathbb{R}^m$ is the desired functional estimate of $u(k)$.

Theorem 4.1 *The qth-order observer (4.17) will estimate $u_f(k)$ if the following conditions hold:*

(i) M is a stable matrix;
(ii) $VT = L - E\bar{C}$;
(iii) $MT + J\bar{C} - T\bar{A} = 0$;
(iv) $H = T\bar{B}$;
(v) $(n + r_d - p) \geq q \geq \rho(L(I_{n+r_d} - \bar{C}^+\bar{C}))$,

where $L \in \mathbb{R}^{r \times (n+r_d)}$ is a known functional gain matrix and $T \in \mathbb{R}^{q \times (n+r_d)}$ is an unknown constant matrix.

Proof Let us define $e(k)$, the error between $\zeta(k)$ and $Tx_d(k)$, as

$$e(k) \triangleq \zeta(k) - Tx_d(k). \tag{4.18}$$

Taking the first-order difference of (4.18) yields

$$e(k+1) = \zeta(k+1) - Tx_d(k+1)$$
$$= Me(k) + (MT + J\bar{C} - T\bar{A})x_d(k) + (H - T\bar{B})u(k) + JGv(k) - T\bar{\Gamma}w(k). \tag{4.19}$$

If conditions (iii) and (iv) of Theorem 4.1 are satisfied, then (4.19) reduces to

$$e(k+1) = Me(k) + JGv(k) - T\bar{\Gamma}w(k). \tag{4.20}$$

The output estimation error is described as

$$e_u(k) \triangleq \hat{u}_f(k) - Lx_d(k) = Ve(k) + (VT + E\bar{C} - L)x_d(k) + EGv(k). \quad (4.21)$$

If condition (ii) of Theorem 4.1 is satisfied, then the above equation reduces to

$$e_u(k) = Ve(k) + EGv(k). \quad (4.22)$$

Hence the output error dynamics $e_u(k)$ is governed by the matrices V and E. The matrix E can be chosen as $E = L\bar{C}^+$ to satisfy the observer condition (v) of Theorem 4.1.

Further, the unknown terms J and T in the observer Eq. (4.17) can be solved as follows:

$$\begin{bmatrix} J & T \end{bmatrix} = \mathscr{X}, \quad (4.23)$$

where $\mathscr{X} \in \mathbb{R}^{q \times (n+p+r_d)}$ is an unknown matrix.

Using (4.23), the matrices J and T can be expressed in terms of the unknown matrix $\mathscr{X}$ as

$$J = \mathscr{X} \begin{bmatrix} I_{p \times p} \\ 0_{(n+r_d) \times p} \end{bmatrix} \quad (4.24)$$

and

$$T = \mathscr{X} \begin{bmatrix} 0_{p \times (n+r_d)} \\ I_{(n+r_d) \times (n+r_d)} \end{bmatrix}. \quad (4.25)$$

On putting the values of J and T in conditions (iii) and (iv) of Theorem 4.1, we obtain

$$M\mathscr{X} \begin{bmatrix} 0 \\ I \end{bmatrix} + \mathscr{X} \begin{bmatrix} \bar{C} \\ -\bar{A} \end{bmatrix} = 0 \quad (4.26)$$

$$V\mathscr{X} \begin{bmatrix} 0 \\ I \end{bmatrix} = L - E\bar{C}. \quad (4.27)$$

We now assemble the matrix Eqs. (4.26) and (4.27) in composite form as follows:

$$\begin{bmatrix} M \\ V \end{bmatrix} \mathscr{X} \begin{bmatrix} 0 \\ I \end{bmatrix} + \begin{bmatrix} I \\ 0 \end{bmatrix} \mathscr{X} \begin{bmatrix} \bar{C} \\ -\bar{A} \end{bmatrix} = \Theta, \quad (4.28)$$

with

$$\Theta = \begin{bmatrix} 0 \\ L - E\bar{C} \end{bmatrix},$$

where $\Theta \in \mathbb{R}^{(q+r) \times (n+r_d)}$ is a known constant matrix.

By applying the Kronecker product [20] in (4.28), we can write

$$\Sigma vec(\mathscr{X}) = vec(\Theta) \tag{4.29}$$

with

$$\Sigma = \left[\begin{bmatrix} 0 \\ I \end{bmatrix}^T \otimes \begin{bmatrix} M \\ V \end{bmatrix} + \begin{bmatrix} \bar{C} \\ -\bar{A} \end{bmatrix}^T \otimes \begin{bmatrix} I \\ 0 \end{bmatrix} \right],$$

where $vec(\mathscr{X}) \in \mathbb{R}^{(n+p)(q+r_d)}$, $vec(\Theta) \in \mathbb{R}^{(n+r_d)(q+r)}$, and $\Sigma \in \mathbb{R}^{(q+r)(n+r_d) \times (q+r_d)(n+p)}$.
The error covariance of (4.20) propagates as

$$P(k+1) = MP(k)M^T + JGRG^T J^T + T\bar{\Gamma}Q\bar{\Gamma}^T T^T. \tag{4.30}$$

The output error covariance of (4.22) is

$$P_u(k) = VP(k)V^T + EGRG^T E^T. \tag{4.31}$$

We can now write $P_u(k+1)$ in terms of the covariance:

$$P_u(k+1) = V(MP(k)M^T + JGRG^T J^T + T\bar{\Gamma}Q\bar{\Gamma}^T T^T)V^T + EGRG^T E^T. \tag{4.32}$$

To minimise the effect of the process and measurement noise covariance, we choose
$\mathscr{X}$ in (4.23) as

$$\mathscr{X} = \underset{\mathscr{X}}{\operatorname{argmin}} \left\| V\mathscr{X} \begin{bmatrix} GRG^T & 0 \\ 0 & \bar{\Gamma}Q\bar{\Gamma}^T \end{bmatrix} \mathscr{X}^T V^T \right\|,$$

thereby minimising the effect of process and measurement noise in the functional
observer. Designing a q-dimensional dynamical system, where

$$(n + r_d - p) \geq q \geq \rho(L(I_{(n+r_d)} - \bar{C}^+\bar{C})),$$

completes the proof.

Based on the above-mentioned development, the design procedure is now summarised as follows:

Algorithm 4.1 Functional Observer-Based SMC with Unmatched Uncertainty

---

1: To obtain the sliding function

$$s(k) \triangleq \bar{c} x_d(k),$$

where $\bar{c} = [c \quad k_0[I_{n-m} \quad 0_{(n-m) \times m}]F]$,

2: Design the controller $u(k)$, and the functional to be estimated is $L x_d(k)$, where $L = [-(cB)^{-1} cA \quad -(cB)^{-1}(c + k_0[I \quad 0])F]$.

3: Choose the $(r \times q)$ elements of the matrix V arbitrarily.

4: To obtain the order of the functional observer

$$(n + r_d - p) \geq q \geq \rho(L(I_{(n+r_d)} - \bar{C}^+ \bar{C}))$$

such that $\rho(V) = \rho(L - E\bar{C})$,

5: Choose arbitrarily a stable $(q \times q)$ matrix M.

6: Solve the composite form of the matrix (4.28) by the Kronecker product. For the matrix $\mathscr{X}$, the matrix $\mathscr{X}$ is chosen according to

$$\mathscr{X} = \underset{\mathscr{X}}{\operatorname{argmin}} \left\| \mathscr{X} \begin{bmatrix} GRG^T & 0 \\ 0 & \bar{\Gamma} Q \bar{\Gamma}^T \end{bmatrix} \mathscr{X}^T \right\|$$

subject to the equality condition (4.28). If satisfied, go to the next step; otherwise, set $q = q + 1$ and go to step 4.

7: From (4.24) and (4.25), find the values of the matrices J and T.

8: Now find the matrix H by $H = T\bar{B}$.

9: As a result of steps 1–8, obtain a structure of the functional observer as in (4.17).

4.5 Simulation Example and Results

In this section, simulation results will be given to demonstrate the validity of the approaches proposed in this chapter. It is assumed that an eighth-order aircraft system is computer controlled [21]. The covariance of process noise and measurement noises are $Q = 0.1I$ and $R = 0.1I$, respectively [22]. Let the unmatched uncertainty $[-15 + 0.2 \sin(0.125\pi k)]$ and matched uncertainty $\sin(k/2) e^{(-k/5)}$ be slowly varying, which affects the discrete-time system. The parameters θ and ε are chosen as $\theta = 0.9$ and $\varepsilon = 0.01$. The system's initial conditions and observer initial conditions are chosen, arbitrarily, as $x_d(0) = [1 \quad 1.5 \quad 2 \quad 2.5 \quad 3 \quad 3.5 \quad 4 \quad 5 \quad 6]^T$ and $\zeta(0) = [1 \quad 2 \quad 7]^T$:

$$A = \begin{bmatrix} 0 & 0 & 1 & 0 & 0 & 0 & 0 & 0 \\ 0 & -0.154 & -0.0042 & 1.54 & 0 & -0.744 & -0.032 & 0 \\ 0 & 0.249 & -1 & -5.2 & 0 & 0.337 & -1.12 & 0 \\ 0.0386 & -0.996 & -0.0003 & -2.117 & 0 & 0.02 & 0 & 0 \\ 0 & 0.5 & 0 & 0 & -4 & 0 & 0 & 0 \\ 0 & 0 & 0 & 0 & 0 & -20 & 0 & 0 \\ 0 & 0 & 0 & 0 & 0 & 0 & -25 & 0 \\ 0 & 1 & 0 & 0 & 0 & 0 & 0 & 1 \end{bmatrix}, B = \begin{bmatrix} 0 & 0 \\ 0 & 0 \\ 0 & 0 \\ 0 & 0 \\ 0 & 0 \\ 20 & 0 \\ 0 & 25 \\ 0 & 0 \end{bmatrix},$$

$$\Gamma = [0\ 0\ 0\ 0\ 0\ 20\ 0\ 0]^T, F = [1\ -2\ 0\ 1\ 0\ 0\ 1\ -0.5]^T.$$

$$C = [I_4 : \mathbf{0}_3], G = [1\ 1\ 1\ 0]^T,$$

where I_4 denotes the 4×4 identity matrix.

The sliding function is computed to be

$$s(k) = \begin{bmatrix} 0.256 & 4.309 & -0.126 & 0.680 & 0.218 & -0.189 & 0.026 & 12.874 & 0.766 \\ -6.852 & -1.053 & -2.877 & 0.801 & -0.102 & -0.075 & 0.237 & -6.085 & -2.292 \end{bmatrix} x_d(k).$$

The control input is obtained as

$$u(k) = \begin{bmatrix} 0.334 & 6.957 & -0.009 & 2.171 & 0.080 & -0.239 & -0.007 & 12.864 & -0.526 \\ 6.694 & -3.165 & 3.677 & -3.156 & 0.037 & -0.044 & -0.163 & 5.962 & 6.067 \end{bmatrix} x_d(k),$$

$$k_0 = \begin{bmatrix} -0.902 & -1.254 & 0.435 & 1.554 & 3.860 & 1.845 \\ -3.437 & -0.001 & -0.005 & 2.099 & -1.234 & 0.002 \end{bmatrix}.$$

We now have the matrix

$$E = \begin{bmatrix} 0.0801 & -0.2390 & -0.0074 & 12.8644 \\ 0.0369 & -0.0438 & -0.1629 & 5.9622 \end{bmatrix}.$$

The order of the functional observer is obtained as $q = 3$.

A diagonal convergent matrix M and arbitrary matrix V are chosen as

$$M = \begin{bmatrix} 0.2 & 0 & 0 \\ 0 & 0.1 & 0 \\ 0 & 0 & 0.3 \end{bmatrix}, V = \begin{bmatrix} 3 & 1 & 4 \\ 5 & 6 & 7 \end{bmatrix}.$$

The matrix $\mathscr{X}$ is obtained as $\mathscr{X} = [\mathscr{X}_1\ \ \mathscr{X}_2]$,
where

$$\mathscr{X}_1 = \begin{bmatrix} 386.5 & 29.4 & 30.3 & -553.2 & -0.1 & -11.8 & 0 & -5.9 & 2302.3 \\ 36.3 & 0.9 & 0.8 & -36.6 & 1.2 & -2.4 & 0.5 & -1.8 & 135.5 \\ -119.5 & -34.3 & -34.9 & 370 & 11.0 & 0.1 & 26.8 & 5.3 & -1760.6 \end{bmatrix}$$

and

$$\mathscr{X}_2 = \begin{bmatrix} -153.4 & -153 & -691.5 & -27.0 \\ -8.9 & -8.9 & -40.6 & 0.1 \\ 117.2 & 117.0 & 528.8 & 20.0 \end{bmatrix}.$$

Solving (4.24) for the matrix J gives

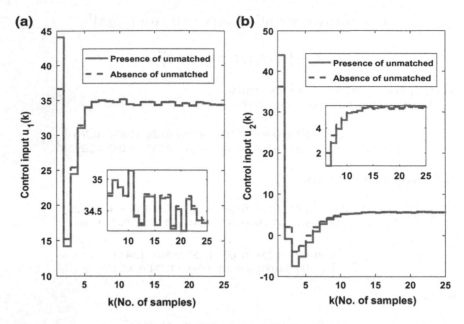

Fig. 4.1 Response for **a** control input $u_1(k)$ and **b** control input $u_2(k)$ in the presence and absence of unmatched uncertainty

$$J = \begin{bmatrix} 386.5059 & 29.4132 & 30.3045 & -553.2156 \\ 36.2853 & 0.8765 & 0.8481 & -36.5774 \\ -119.5077 & -34.2675 & -34.8614 & 370.1483 \end{bmatrix},$$

and solving (4.25) for the matrix T gives

$$T = \begin{bmatrix} -0.1 & -11.8 & 0 & -5.9 & 2302.3 & -153.4 & -153 & -691.5 & -27.0 \\ 1.2 & -2.4 & 0.5 & -1.8 & 135.5 & -8.9 & -8.9 & -40.6 & 0.1 \\ -0.1 & 11.0 & 0.1 & 5.3 & -1760.6 & 117.2 & 117.0 & 528.8 & 20.0 \end{bmatrix}.$$

Since $H = TB$, the matrix H is obtained as

$$H = \begin{bmatrix} -153.6423 & -152.7872 \\ -8.6993 & -9.0225 \\ 117.1614 & 116.7549 \end{bmatrix}.$$

Then the estimate of $u(k)$ is given by the functional observer (4.17).

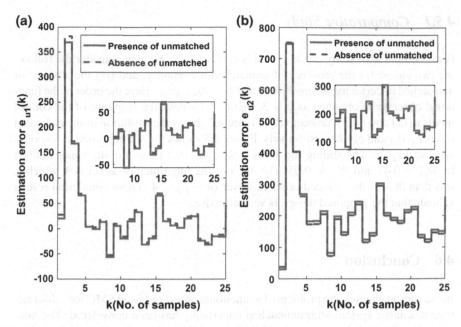

Fig. 4.2 Response for **a** estimation error $e_{u1}(k)$ and **b** estimation error $e_{u2}(k)$ in the presence and absence of unmatched uncertainty

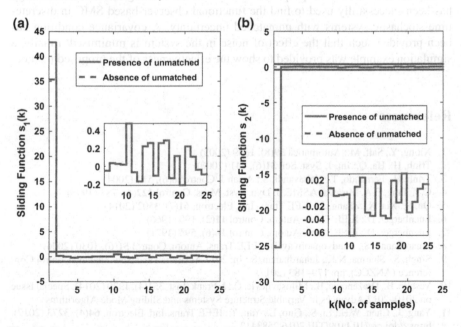

Fig. 4.3 Response for **a** the sliding function $s_1(k)$ and **b** the sliding function $s_2(k)$ in the presence and absence of unmatched uncertainty

4.5.1 Comparative Study

For a comparative study of the system's performance, we can consider the following two cases: (i) the presence of unmatched uncertainty, and (ii) the absence of unmatched uncertainty, where $F_1 = 0$ and $k_0 = 0_{m \times (n-m)}$. Here the order of the functional observer is obtained as $q = 3$. Figure 4.1 shows the response of the control inputs $u_1(k)$ and $u_2(k)$, respectively. Figure 4.2 shows the evolution of the estimated errors $e_{u1}(k)$ and $e_{u2}(k)$, respectively. Figure 4.3 shows the sliding function response $s_1(k)$ and $s_2(k)$. The sliding functions $s_1(k)$ and $s_2(k)$ lie within the bands defined by $W_{c1} = 0.47$ and $W_{c2} = 0.05$. The order of the functional observer is 3, which is less than that of the reduced-order observer $(n - p) = 4$. These simulation results validate that the proposed design is very effective.

4.6 Conclusion

In the current chapter, the problem of a functional observer-based SMC for a discrete-time stochastic system with unmatched uncertainty has been considered. The minimisation of unmatched uncertainty effects in a sliding mode, disturbance-dependent sliding surface method was proposed. Along with the existing conditions, a strategy has been successfully used to find the functional observer-based SMC in discrete-time stochastic systems with unmatched uncertainty. A covariance condition has been provided such that the effect of noise in the system is minimised. Finally, a simulation example was provided to show the effectiveness of the proposed method.

References

1. Xiong, Y., Saif, M.: Automatica **39**(8), 1389 (2003)
2. Trinh, H., Ha, Q.: Int. J. Syst. Sci. **31**(6), 741 (2000)
3. Trinh, H., Fernando, T., Nahavandi, S.: Asian J. Control **6**(6), 514 (2004)
4. C.Y. Tang, E.A. Misawa, ASME, J. Dyn. Syst. Meas. Control **122**(4), 783 (1998)
5. Qu, S., Xia, X., Zhang, J.: IEEE Trans. Ind. Electron. **61**(7), 3502 (2014)
6. Luenberger, D.: IEEE Trans. Autom. Control **11**(2), 190 (1966)
7. Luenberger, D.: IEEE Trans. Autom. Control **16**(6), 596 (1971)
8. Janardhanan, S., Bandyopadhyay, B.: IEEE Trans. Autom. Control **51**(6), 1030 (2006)
9. Singh, S., Sharma, N.K., Janardhanan, S.: In: 2017 Australian and New Zealand Control Conference (ANZCC), pp. 179–183 (2017)
10. Veselić, B., Draženović, B., Milosavljević, Č.: J. Frank. Inst. **351**(4), 1920 (2014). Special Issue on 2010–2012 Advances in Variable Structure Systems and Sliding Mode Algorithms
11. Yang, J., Chen, W.H., Li, S., Guo, L., Yan, Y.: IEEE Trans. Ind. Electron. **64**(4), 3273 (2017). https://doi.org/10.1109/TIE.2016.2583412
12. Monsees, G., Scherpen, J.M.A.: In: 2001 European Control Conference (ECC), pp. 3270–3275 (2001)
13. Polyakov, A., Poznyak, A.: Automatica **47**(7), 1450 (2011)

14. Ma, J., Ni, S., Xie, W., Dong, W.: In: 2015 IEEE International Conference on Information and Automation, pp. 2930–2936 (2015)
15. Xi, Z., Hesketh, T.: IET Control Theory Appl. **4**(5), 889 (2010)
16. Xi, Z., Hesketh, T.: IET Control Theory Appl. **4**(10), 2071 (2010)
17. Singh, S., Janardhanan, S.: Int. J. Syst. Sci. **50**(6), 1179 (2019). https://doi.org/10.1080/00207721.2019.1597942
18. Koshkouei, A.J., Zinober, A.S.I.: ASME **122**(4), 793 (2000). https://doi.org/10.1115/1.1321266
19. Janardhanan, S., Kariwala, V.: IEEE Trans. Autom. Control **53**(1), 367 (2008)
20. Brewer, J.: IEEE Trans. Circuits Syst. **25**(9), 772 (1978)
21. Edwards, C., Spurgeon, S.: Sliding Mode Control: Theory And Applications. Series in Systems and Control. Taylor & Francis, Milton Park (1998)
22. Sage, A., Mels, J.: Estimation Theory with Applications to Communications and Control. McGraw-Hill Series in Systems Science. McGraw-Hill, New York (1971)

References

Chapter 5
Stochastic Sliding Mode Control for Parametric Uncertain Systems Using a Functional Observer

Abstract This chapter addresses the problem of sliding mode control (SMC) design using functional observers (FO) for discrete-time stochastic systems with norm-bounded parametric uncertainty in the system matrix. SMC is a powerful, robust control technique to stabilise discrete-time systems. However, the application of SMC becomes difficult when the system states are inaccessible for feedback. Hence the proposed method uses an FO-based approach for estimating SMC design. To mitigate the side effect of the parametric uncertainty on the estimation error, a sufficient condition for stability is provided based on Gershgorin's circle theorem. The feasibility and effectiveness of our derived results are illustrated through a simulation example.

Keywords Discrete-time sliding mode control · Stochastic systems · Parametric uncertainty · Functional observers

5.1 Introduction

In Chaps. 3 and 4, functional observer-based SMCs for systems without and with unmatched uncertainty have been designed. In [1–3], the authors developed functional observers for perturbed dynamical systems with disturbance-reduced effects on the estimation errors, but they do not consider the presence of parametric uncertainties. However, parametric uncertainties are always present in the system matrix, which was not discussed in the previous chapter. Designing functional observers without taking into account these robustness constraints may destabilise the estimation errors, causing a discrepancy between the states of the system and the estimates of those states [4], which means that the estimation error should converge asymptotically to zero in the absence of disturbances [5, 6]. At the same time, when the system matrices are affected by parametric uncertainties, the functional observer must minimise the effect of the system's states on the estimated error. Despite the importance of estimation of system states under certain parameters due to modelling errors, to the best of the authors' observation, there is no work in which the problem of designing functional observer-based SMC for stochastic systems is investigated.

© Springer Nature Switzerland AG 2020

S. Singh and S. Janardhanan, *Discrete-Time Stochastic Sliding Mode Control Using Functional Observation*, Lecture Notes in Control and Information Sciences 483, https://doi.org/10.1007/978-3-030-32800-9_5

Inspired by the preceding discussion, this chapter extends the results of Chap. 3 to the case of parametric uncertainty present in the system state matrix. Here, a sliding function is designed by the linear matrix inequalities (LMI) approach. Efficient convex optimisation techniques are available to solve LMI problems [7].

The main contributions of this chapter are threefold:

1. The sliding function is investigated for parametric uncertain discrete-time stochastic systems.
2. The functional observer-based SMC is designed for parametric uncertain discrete-time stochastic systems.
3. To mitigate the side effect of the parametric uncertainty on the estimation error, a sufficient condition on stability is proposed based on Gershgorin's circle theorem.

The subsequent sections are organised as follows. In Sect. 5.2, the problem of functional observer-based SMC design is formulated. In Sect. 5.3, a sliding function and SMC are designed in the presence of parametric uncertainty. The functional observer-based SMC design along with existence conditions and design analysis are presented in Sect. 5.4. Efficiency of the proposed DSMC is studied by numerical example in Sect. 5.5. Finally, Sect. 5.6 reviews the contributions made in this chapter.

5.2 Problem Formulation

Consider the parametric uncertain discrete-time stochastic system [8]

$$\begin{aligned}
x(k+1) &= (A + \Delta A)x(k) + Bu(k) + \Gamma w(k), \\
y(k) &= Cx(k) + Gv(k),
\end{aligned} \tag{5.1}$$

where $x(k) \in \mathbb{R}^n$, $u(k) \in \mathbb{R}^m$ and $y(k) \in \mathbb{R}^p$, with matrices of appropriate dimensions; ΔA is the parametric uncertainty term for the system matrix. The covariance of the plant noise is Q, and the covariance of the measurement noise is R. We assume that the initial state $x(0)$ is a Gaussian random vector, and $x(0)$, $w(k)$, and $v(k)$ are mutually uncorrelated. The following assumptions are needed throughout this chapter.

Assumption 5.1 The system (5.1) is completely controllable and observable.

Assumption 5.2 The input uncertainty satisfies the matching condition [9], i.e., $rank[B, \Gamma] = rank[B]$ but ΔA does not lie in the range space of B and hence is mismatched.

Assumption 5.3 The system parametric uncertainties are norm-bounded with the form $\Delta A = G_1 \Delta(k) E_1$, where $G_1 \in \mathbb{R}^{n \times n_a}$ and $E_1 \in \mathbb{R}^{n_b \times n}$ are known real constant matrices, and the properly dimensioned matrix $\Delta(k) \in \mathbb{R}^{n_a \times n_b}$ is unknown but norm-bounded as $\Delta^T(k)\Delta(k) \leq I$, where I denotes the identity matrix with appropriate dimension.

The main goal is to design a functional observer-based SMC law for the system (5.1) such that the sliding function $s(k)$ is stabilised in a specified band and the stochastic sliding mode, as defined in (1.18), can be achieved.

5.3 Sliding Function and Controller Design

5.3.1 Sliding Function Design

As defined in (1.9), the sliding function analysis is the same as for the uncertain system (3.1).

Assumption 5.4 cB is a nonsingular matrix.

5.3.1.1 Stability Analysis

To assist the progress of the following DSMC design, a state transformation $\xi(k) = Ux(k)$ is introduced, where $U \in \mathbb{R}^{n \times n}$ is a nonsingular matrix satisfying

$$UB = \begin{bmatrix} 0_{(n-m) \times m} \\ B_2 \end{bmatrix}, \tag{5.2}$$

where $B_2 \in \mathbb{R}^{m \times m}$ is nonsingular. Then the system (5.1) can be transformed into regular form:

$$\xi_1(k+1) = (\bar{A}_{11} + \Delta\bar{A}_{11})\xi_1(k) + (\bar{A}_{12} + \Delta\bar{A}_{12})\xi_2(k),$$
$$\xi_2(k+1) = (\bar{A}_{21} + \Delta\bar{A}_{21})\xi_1(k) + (\bar{A}_{22} + \Delta\bar{A}_{22})\xi_2(k) + B_2u(k) + \Gamma_2w(k),$$
$$\tag{5.3}$$

where

$$U(A + \Delta A)U^{-1} = \begin{bmatrix} \bar{A}_{11} + \Delta\bar{A}_{11} & \bar{A}_{12} + \Delta\bar{A}_{12} \\ \bar{A}_{21} + \Delta\bar{A}_{21} & \bar{A}_{22} + \Delta\bar{A}_{22} \end{bmatrix}, \quad and \quad U\Gamma = \begin{bmatrix} 0 \\ \Gamma_2 \end{bmatrix}.$$

The sliding function (1.9) is expressed in terms of the new state $\xi(k)$ as

$$s(k) \triangleq cU^T\xi(k) \triangleq K\xi_1(k) + \xi_2(k), \tag{5.4}$$

where the gain matrix is $K \in \mathbb{R}^{m \times (n-m)}$. Substituting $\xi_2(k) = -K\xi_1(k)$ in the first equation of system (5.3) gives

$$\xi_1(k+1) = (\bar{A}_{11} + \Delta\bar{A}_{11} - \bar{A}_{12}K - \Delta\bar{A}_{12}K)\xi_1(k). \tag{5.5}$$

The objective is to design a gain matrix K such that the system (5.5) is quadratically stable. Before proceeding further, a lemma that plays an important role in the derivation of the sliding function design is introduced. To this end, the following lemmas are necessary.

Lemma 5.1 ([10, 11]) *Given matrices* $\Sigma_1 = \Sigma_1^T$, Σ_2, *and* Σ_3 *with the appropriate dimensions, the inequality*

$$\Sigma_1 + \Sigma_2 \Delta(k)\Sigma_3 + \Sigma_3^T \Delta^T(k)\Sigma_2^T < 0$$

holds $\forall \Delta(k)$ *satisfying* $\Delta^T(k)\Delta(k) \leq I$ *if and only if there exists a scalar* $\lambda > 0$ *such that*

$$\Sigma_1 + \lambda^{-1}\Sigma_2\Sigma_2 + \lambda\Sigma_3^T \Sigma_3^T < 0.$$

Lemma 5.2 ([10]) *For a real symmetric matrix* $\Omega = \begin{bmatrix} \Omega_{11} & \Omega_{12} \\ \Omega_{12}^T & \Omega_{22} \end{bmatrix}$, *the following three conditions are equivalent:*

(i) $\Omega < 0$;
(ii) $\Omega_{11} < 0$ *and* $\Omega_{22} - \Omega_{12}^T\Omega_{11}^{-1}\Omega_{12} < 0$;
(iii) $\Omega_{22} < 0$ *and* $\Omega_{11} - \Omega_{12}\Omega_{22}^{-1}\Omega_{12}^T < 0$.

The primary results regarding the design of sliding surfaces is summarised in Theorem 5.1.

Theorem 5.1 *The reduced-order uncertain system (5.5) is quadratically stable if there exist symmetric positive definite matrices* $Z \in \mathbb{R}^{(n-m)\times(n-m)}$, *a matrix* $Y \in \mathbb{R}^{m\times(n-m)}$, *and a scalar* $\lambda > 0$ *such that the following LMI holds:*

$$\begin{bmatrix} -Z & * & * & * \\ \bar{A}_{11}Z - \bar{A}_{12}Y & -Z & * & * \\ E_{11}Z - E_{12}Y & 0 & -\lambda I & * \\ 0 & \lambda G_1^T & 0 & -\lambda I \end{bmatrix} < 0. \qquad (5.6)$$

Moreover, the gain matrix K *is given by* $K = YZ^{-1}$.

Proof For the system in (5.5), define a Lyapunov function as $V(k) = \xi_1^T(k)Z^{-1}\xi_1(k)$, which is positive $\forall \xi_1(k) \neq 0$. For stability, the finite difference $\Delta V(k)$ of the Lyapunov function must be such that

$$\Delta V(k) = V(k+1) - V(k)$$
$$= \xi_1^T(k)[\tilde{A}_{11}^T Z^{-1}\tilde{A}_{11} - Z^{-1}]\xi_1(k) < 0, \qquad (5.7)$$

where $\tilde{A}_{11} = (\bar{A}_{11} + \Delta\bar{A}_{11} - \bar{A}_{12}K - \Delta\bar{A}_{12}K)$.

Using Lemma 5.2, the aforementioned equation can be transformed into

$$\begin{bmatrix} -Z^{-1} & \tilde{A}_{11}^T \\ \tilde{A}_{11} & -Z \end{bmatrix} < 0. \tag{5.8}$$

The inequality (5.8) can be rewritten as

$$\begin{bmatrix} -Z^{-1} & * \\ (\bar{A}_{11} - \bar{A}_{12}K) & -Z \end{bmatrix} + \begin{bmatrix} 0 \\ G_1 \end{bmatrix} \Delta(k) [E_{11} - E_{12}K \ 0]$$

$$+ [E_{11} - E_{12}K \ 0]^T \Delta^T(k) \begin{bmatrix} 0 \\ G_1 \end{bmatrix}^T < 0. \tag{5.9}$$

In light of Lemma 5.1, the inequality (5.9) holds for every $\Delta^T(k)$ satisfying $\Delta^T(k)\Delta(k) \le I$ if and only if there exists a scalar $\lambda > 0$ such that

$$\begin{bmatrix} -Z^{-1} & * \\ (\bar{A}_{11} - \bar{A}_{12}K) & -Z \end{bmatrix} + \lambda \begin{bmatrix} 0 \\ G_1 \end{bmatrix} [0 \ G_1^T] + \lambda^{-1} \begin{bmatrix} E_{11}^T - K^T E_{12}^T \\ 0 \end{bmatrix} [E_{11} - E_{12}K \ 0] < 0. \tag{5.10}$$

Applying Schur's complement lemma, Lemma 5.2, Eq. (5.10) can be rewritten as

$$\begin{bmatrix} -Z^{-1} & * & * & * \\ \bar{A}_{11} - \bar{A}_{12}K & -Z & * & * \\ E_{11} - E_{12}K & 0 & -\lambda I & * \\ 0 & G_1^T & 0 & -\lambda^{-1}I \end{bmatrix} < 0. \tag{5.11}$$

Taking the congruence transformation by the matrix (5.11) $diag\{Z, I_{n-m}, I_{n-m}, I_{n-m}\}$ yields

$$\begin{bmatrix} -Z & * & * & * \\ \bar{A}_{11}Z - \bar{A}_{12}KZ & -Z & * & * \\ E_{11}Z - E_{12}KZ & 0 & -\lambda I & * \\ 0 & G_1^T & 0 & -\lambda^{-1}I \end{bmatrix} < 0. \tag{5.12}$$

Setting $Y = KZ$ and taking the congruence transformation with $diag\{I, I, I, \lambda I\}$ yields (5.6). It can be easily shown that $\Delta V(k) < 0$ if the LMI (5.6) is satisfied. Therefore, the reduced-order system (5.5) is quadratically stable with $K = YZ^{-1}$.

Moreover, the quadratically stable sliding surface of (5.4) is

$$s(k) = YZ^{-1}\xi_1(k) + \xi_2(k) = 0.$$

This completes the proof.

5.3.2 Design of Controller

Our main goal is to find out the control that achieves the same objective as in (1.18) for the system (5.1).

The bounds of the disturbance vector are known; let [9]

$$d_l \le c\Delta Ax(k) \le d_u,$$

where the lower bound d_l and the upper bound d_u are known constants. Furthermore, the average value of the disturbance d_0 is $(d_l + d_u)/2$.

The SMC can be obtained as

$$u(k) = -(cB)^{-1}(cAx(k) + d_0), \tag{5.13}$$

and the variance is $\sigma_c^2 = c\Gamma Q\Gamma^T c^T$. The rest of the analysis is similar to that of Chap. 1. Hence it is omitted here. When the SMC (5.13) is used, the sliding variable $s(k)$ will have the dynamics

$$s(k+1) = c\Gamma w(k) + c\Delta Ax(k) - d_0.$$

Then the system motion (5.1) is called stochastic sliding mode for uncertain parametric systems.

5.4 Design of Functional Observer-Based SMC

The structure of the functional observer is the same as in (3.8) for system (5.1). Similar to the discussion in Sect. 3.4, the estimate of $u(k) = Lx(k)$, using a qth-order observer given (3.8) for the system (5.1), can be obtained, provided the condition stated in Theorem 3.1 holds. Due to the presence of parametric uncertainty in system the dynamics (5.1), the error dynamics of the error $e(k) \triangleq \zeta(k) - Tx(k)$ will be different from that of (3.13) and can be obtained as

$$
\begin{aligned}
e(k+1) &= \zeta(k+1) - Tx(k+1) \\
&= M\zeta(k) + Jy(k) + Hu(k) - T(A + \Delta A)x(k) - TBu(k) - T\Gamma w(k) \\
&= Me(k) + (MT + JC - TA)x(k) - T\Delta Ax(k) + (H - TB)u(k) + JGv(k) \\
&\quad - T\Gamma w(k).
\end{aligned}
\tag{5.14}
$$

Applying conditions (ii) and (iii) of Theorem 3.1 to Eq. (5.14) yields

$$e(k+1) = Me(k) - T\Delta Ax(k) + JGv(k) - T\Gamma w(k). \tag{5.15}$$

Hence the dynamics of the error $e(k)$ is governed by the matrices M, J, and T. On the other hand, to mitigate the side effect of the parametric uncertainties on the dynamics of the estimation error (5.15), the following definition must be satisfied.

Definition 5.4.1 Gershgorin's Circle Theorem [12]
Let $\mathscr{A} = [a_{ij}] \in \mathbb{C}^{n \times n}$ be a matrix, and let r_i denote the sum of the absolute values of the off-diagonal entries in the ith row of $\mathscr{A}_i$, that is, $r_i = \Sigma_{j=1, j \neq i}^n |a_{ij}|$. The ith Gershgorin disc is the circular disc $\mathscr{D}_i$ in the complex plane with centre a_{ii} and radius r_i. That is,

$$\mathscr{D}_i = \{\mathscr{Z} \in \mathbb{C}_i : |\mathscr{Z} - a_{ii}| \leq r_i\},$$

where r_i denotes the deleted absolute row sums of $\mathscr{A}$, i.e.,

$$r_i = \Sigma_{j \neq i}^n |a_{ij}|. \tag{5.16}$$

Then all the eigenvalues of $\mathscr{A}_i$ lie in the union of the discs $\mathscr{D}_i$ for $i = 1, ..., n$.

Gershgorin's circle theorem states that the eigenvalues of a matrix are contained within a union of a series of discs in the complex plane.

Augment (5.1) and (5.15) as a matrix form

$$\mathbb{G}(k+1) = \mathscr{K}\mathbb{G}(k) + \mathscr{B}w(k) + \mathscr{B}_1 v(k), \tag{5.17}$$

where $\mathbb{G}(k) = [x^T(k) \quad e^T(k)]^T \in \mathbb{R}^{(n+q)}$,

$$\mathscr{K} = \begin{bmatrix} A + BL + G_1 \Delta(k) E_1 & 0 \\ -T \Delta A & M \end{bmatrix}, \mathscr{B} = \begin{bmatrix} \Gamma \\ -T \Gamma \end{bmatrix}, \mathscr{B}_1 = \begin{bmatrix} 0 \\ JG \end{bmatrix}.$$

Using Gershgorin's circle theorem, if each eigenvalue of $\mathscr{K}$ lies in the union of the circles, then the system will be stable. For this method, we obtain the bound $|r_i(\mathscr{K})| + |a_{ii}| < 1$. In that case, the following condition holds: $|\mathscr{Z} - \mathscr{K}_{ii}| \leq r_i(\mathscr{K})$.

We define the output estimation error as follows:

$$e_u(k) \triangleq \hat{u}_f(k) - Lx(k) = Ve(k) + (VT + EC - L)x(k) + EGv(k). \tag{5.18}$$

Applying condition (iv) of Theorem 3.1 to Eq. (5.18) yields

$$e_u(k) = Ve(k) + EGv(k). \tag{5.19}$$

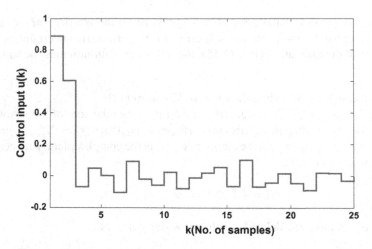

Fig. 5.1 Evolution of control input $u(k)$

Hence the output error (5.19) is governed by the matrices V and E. The solutions of the matrices E, J, and T are obtained as given in Sect. 3.4 of Chap. 3. This completes the proof.

The procedure of the proposed DSMC using functional observers for uncertain stochastic systems are summarised as follows:

Algorithm 5.1 Design procedure of DSMC using Functional Observer

---

1: To solve the LMI problem (5.6), and obtain K to construct the sliding gain c as in (5.4), and then find the value of L.

2: Obtain the minimum order of the functional observer

$$(n - p) \geq q \geq \rho(L(I_n - C^+C))$$

such that $rank(V) = rank(L - EC)$

3: Choose, Schur matrix M and arbitrarily matrix V.

4: Solve composite form of matrix as in (3.22) by Kronecker product, For matrix $\mathscr{X}$.

5: From (3.18) and (3.19) as in given, Find the values of matrix J and T.

6: As a result of steps 1 to 5 above, obtain a structure of functional observer by (3.8).

5.5   Simulation Results

A numerical example of an uninterruptible power system (UPS) is considered for simulation studies. Consider the model of the UPS as given in [13] in the form of (5.1) with parameters

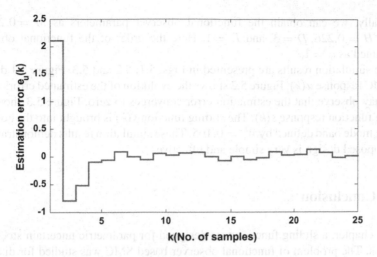

Fig. 5.2 Evolution of estimation error $e_u(k)$

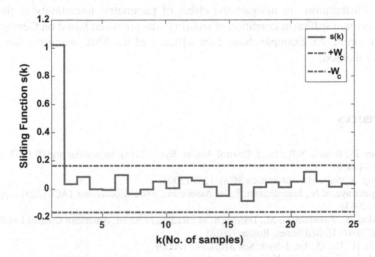

Fig. 5.3 Evolution of sliding function $s(k)$

$$A = \begin{bmatrix} 0.9226 & -0.6330 & 0 \\ 1 & 0 & 0 \\ 0 & 1 & 0 \end{bmatrix}, B = \begin{bmatrix} 1 \\ 0 \\ 0 \end{bmatrix}, \Gamma = \begin{bmatrix} 1 \\ 0 \\ 0 \end{bmatrix}, G_1 = \begin{bmatrix} 0.05 \\ 0.015 \\ 0.08 \end{bmatrix}$$

$$E_1 = \begin{bmatrix} -0.05 & 0.06 & 0.10 \end{bmatrix}, \Delta(k) = 0.3 sin(k), C = \begin{bmatrix} 0 & 0 & 1 \end{bmatrix}, G = 1.$$

The covariances of the process noise and measurement noises are $Q = 0.1I$ and $R = 0.1I$, respectively. The parameters θ and ε are given as $\theta = 0.9$ and $\varepsilon = 0.01$ [14, Remark 2.2]. The primary conditions of the system state and observer state are selected as $x(0) = [1 \quad 0 \quad 0]^T$ and $\zeta(0) = 2$.

Finally, we can obtain the functional observer parameters as $M = 0.2, J = 0.002, H = 0.226, D = 3$, and $E = 1$. Here the order of the functional observer is obtained as $q = 1$.

The simulation results are presented in Figs. 5.1, 5.2 and 5.3. Figure 5.1 depicts the SMC response $u(k)$. Figure 5.2 shows the evolution of the estimated errors $e_u(k)$. One may observe that the estimation error converges to zero. Figure 5.3 shows the sliding function response $s(k)$. The sliding function $s(k)$ is brought into the constant sliding mode band defined by $W_c = 0.165$. These simulation results demonstrate that the proposed design is very simple and effective.

5.6 Conclusion

In this chapter, a sliding function was designed for parametric uncertain stochastic systems. The problem of functional observer-based SMC was studied for discrete-time stochastic systems in the presence of parametric uncertainty in the system matrix. Furthermore, to mitigate the effect of parametric uncertainty in the estimation error, a sufficient condition of stability was proposed based on Gershgorin's circle theorem. An example showed the efficacy of the SMC using the functional observer method.

References

1. Aloui, R., Braiek, N.B.: Int. J. Control Autom. Syst. (2018). https://doi.org/10.1007/s12555-017-0144-9
2. Xiong, Y., Saif, M.: Automatica **39**(8), 1389 (2003)
3. Satyanarayana, N., Janardhanan, S.: In: American Control Conference (ACC), 2014, pp. 451–456 (2014)
4. Boukal, Y., Zasadzinski, M., Darouach, M., Radhy, N.E.: In: American Control Conference, ACC'2016. United States, Boston (2016)
5. Trinh, H., Ha, Q.: Int. J. Syst. Sci. **31**(6), 741 (2000)
6. Trinh, H., Fernando, T., Nahavandi, S.: Asian J. Control **6**(6), 514 (2004)
7. Scherer, C., Weiland, S.: Linear Matrix Inequalities in Control. The Control Systems Handbook, 2nd edn. (Electrical Engineering Handbook) (2010)
8. Singh, S., Janardhanan, S.: IEEE Trans. Circuits Syst. II: Express Briefs **66**(8), 1346 (2019). https://doi.org/10.1109/TCSII.2018.2878973
9. Bandyopadhyay, B., Janardhanan, S.: Discrete-time Sliding Mode Control: A Multirate Output Feedback Approach. Lecture Notes in Control and Information Sciences. Springer, Berlin (2005)
10. Xia, Y., Jia, Y.: IEEE Trans. Autom. Control **48**(6), 1086 (2003)
11. Lin, Y., Shi, Y., Burton, R.: IEEE/ASME Trans. Mechatron. **18**(1), 1 (2013). https://doi.org/10.1109/TMECH.2011.2160959
12. Meyer, C.D. (ed.): Matrix Anal. Appl. Linear Algebra. Society for Industrial and Applied Mathematics, Philadelphia, PA, USA (2000)
13. Argha, A., Li, L., Su, S.W., Nguyen, H.: J. Frankl. Inst. **353**(15), 3857 (2016). https://doi.org/10.1016/j.jfranklin.2016.06.018
14. Singh, S., Janardhanan, S.: Int. J. Syst. Sci. **48**(15), 3246 (2017)

Chapter 6
Functional Observer-Based Sliding Mode Control for Discrete-Time Delayed Stochastic Systems

Abstract This chapter addresses the problem of stabilisation, observer-based sliding mode control, and functional observer-based sliding mode control. An SMC method is proposed for discrete-time delayed stochastic systems. Stability and convergence analyses of the proposed method are provided. Furthermore, the DSMC of a delayed stochastic system for incomplete state information has also been considered, where states are estimated by the Kalman filter approach. A functional observer-based SMC method for discrete-time delayed stochastic systems is proposed. Therefore, the SMC has been estimated by the functional observer approach. Finally, functional observer-based state feedback and the SMC law are compared graphically as well as numerically.

Keywords Discrete-time systems · Stochastic system · State delay · Sliding mode control · State estimation · Linear matrix inequalities · Kalman filter · Linear functional observer

6.1 Introduction

As noted in Chaps. 2–5, proposed SMC design techniques for stochastic systems have been without time delays. However, time delays are a common phenomenon in practical applications and often cause system instability [1–4]. The stabilisation analysis of discrete-time systems with delays is more challenging than that of systems without delays [5]. Therefore, SMC design for discrete-time delayed systems has received attention in recent years [6–10]. But in the background of stochastic DSMC, few works in the literature are available for time-delayed systems [11–13].

All the aforementioned stabilisation techniques are based on full state information, and thus do not apply to the situation in which the state variables are inaccessible for feedback measurement. For this, the Kalman filter approach is one of the most popular techniques for dealing with the state estimation problem for stochastic systems [14–18]. The main advantage of the combination of SMC and Kalman filter is that it eliminates the need for measurability of states for implementation of SMC for stochastic systems. Further, the observer output feedback design for discrete-time

© Springer Nature Switzerland AG 2020

S. Singh and S. Janardhanan, *Discrete-Time Stochastic Sliding Mode Control Using Functional Observation*, Lecture Notes in Control and Information Sciences 483, https://doi.org/10.1007/978-3-030-32800-9_6

systems with delay is investigated in [19–21]. Exploitation of functional observers may increase the applicability of the aforementioned output feedback problems. In related works on discrete-time functional observer design, Koshkouei and Zinober [22] considered systems subject to a stochastic system without delays, while systems with constant delays and no stochastic system are studied in [23–26]. However, research on DSMC for delayed state stochastic systems with stabilisation, observer, and functional observer has not yet been undertaken, which constitutes our research motivation.

In this chapter, we investigate SMC for time-delay stochastic systems with and without states information such that stability of the sliding mode and the system is obtained. Furthermore, this chapter explores the problem of designing functional observer-based SMC for discrete-time stochastic systems with time delay.

Following are the major contributions of this chapter:

1. The sliding function and SMC have been designed for discrete-time delayed stochastic systems. Sliding function stability analysis has been done by the linear matrix inequalities (LMI) approach.
2. Furthermore, DSMC of a time-delayed stochastic system for incomplete state information has also been considered, where states are estimated by the Kalman filter approach.
3. Moreover, SMC has been calculated by the functional observer approach.
4. Finally, the functional observer-based state feedback and SMC law are compared graphically as well as numerically.

The remainder of this chapter is organised as follows. After this introduction, the problem is formulated in Sect. 6.2. An SMC is designed for the complete state information case in Sect. 6.3. In Sect. 6.4, an SMC is designed for the incomplete state information case. A functional observer-based SMC design is presented in Sect. 6.5. Section 6.6 is devoted to a simulation example that highlights the performance of the proposed scheme. Section 6.7 discusses contributions made in this chapter and concludes it.

6.2 System Description and Problem Formulation

Consider a constant delayed discrete-time stochastic system:

$$x(k + 1) = Ax(k) + A_d x(k - k_x) + Bu(k) + \Gamma w(k),$$
$$x(k) = \phi(k), k \in [-k_x, \quad 0], \tag{6.1}$$

with observation equation of the form

$$y(k) = Cx(k) + Gv(k), \tag{6.2}$$

where $x(k)$, $x(k - k_x)$, $u(k)$, and $y(k)$ with the matrices being of appropriate dimensions. The matrices A, $A_d \in \mathbb{R}^{n \times n}$, $B \in \mathbb{R}^{n \times m}$, $C \in \mathbb{R}^{p \times n}$, $G \in \mathbb{R}^{p \times l}$, and $\Gamma \in \mathbb{R}^{n \times \bar{r}}$ are constant and known. Both the state noise sequence $w(k) \in \mathbb{R}^{\bar{r}}$ and the output noise sequence $v(k) \in \mathbb{R}^{l}$ are assumed to be mean-zero white Gaussian noise with covariance Q and R, respectively, with Q nonnegative definite and R positive definite.

The sliding function is similar to that defined in (1.9) for complete state information, and the sliding function is similar to defined in (1.48) for incomplete state information.

The following assumptions will hold throughout this chapter.

Assumption 6.1 The matching condition $rank(B) = rank(B|\Gamma)$ is satisfied [27, 28].

Assumption 6.2 The pair (A, B) is controllable, and the pair (A, C) is observable.

Assumption 6.3 The initial state x_0 and x_{-k_x} are Gaussian random variables, independent of w_k and v_k. Further, x_0 and x_{-k_x} are correlated.

Assumption 6.4 cB is a nonsingular matrix.

The vital aim is to design an SMC using a functional observer for systems (6.1) such that the sliding mode is achieved.

Three types of SMC design problems are considered, comprising the following cases:

(i) Complete state information case,
(ii) Incomplete state information case,
(iii) Functional observer-based based SMC.

6.3 Design of DSMC for Stochastic Systems with Complete State Information

In this section, the design of DSMC for delayed systems is presented. DSMC consists of two steps. First, a sliding function is designed using the LMI method such that the system's responses in the sliding band act like the desired dynamics. Second, an SMC law is obtained such that the sliding mode is reached and remained in for all time.

6.3.1 Sliding Function Design

Consider the matrix U as defined in (5.2) such that $UB = \begin{bmatrix} 0_{(n-m) \times m} \\ B_2 \end{bmatrix}$ and one has the following partitioned matrices:

$$UAU^{-1} = \begin{bmatrix} \bar{A}_{11} & \bar{A}_{12} \\ \bar{A}_{21} & \bar{A}_{22} \end{bmatrix}, \quad UA_dU^{-1} = \begin{bmatrix} \bar{A}_{11} & \bar{A}_{12} \\ \bar{A}_{21} & \bar{A}_{22} \end{bmatrix}, \quad \text{and} \quad U\Gamma = \begin{bmatrix} 0 \\ \Gamma_2 \end{bmatrix}.$$

Thus, the original system (6.1) can be transformed into the regular form

$$\xi_1(k+1) = \bar{A}_{11}\xi_1(k) + \bar{A}_{d11}\xi_1(k-k_x) + \bar{A}_{12}\xi_2(k) + \bar{A}_{d12}\xi_2(k-k_x)$$
$$\xi_2(k+1) = \bar{A}_{21}\xi_1(k) + \bar{A}_{d21}\xi_1(k-k_x) + \bar{A}_{22}\xi_2(k) + \bar{A}_{d22}\xi_2(k-k_x) + B_2u(k)$$
$$+ \Gamma_2 w(k).$$

$$(6.3)$$

The sliding function (1.9) can be expressed as

$$s(k) \triangleq cU^T\xi(k) \triangleq [K \quad I_m]\xi(k) = K\xi_1(k) + \xi_2(k). \tag{6.4}$$

Substituting $\xi_2(k) = -K\xi_1(k)$ into the first equation of the system (6.3) gives

$$\xi_1(k+1) = (\bar{A}_{11} - \bar{A}_{12}K)\xi_1(k) + (\bar{A}_{d11} - \bar{A}_{d12}K)\xi_1(k-k_x), \tag{6.5}$$

where the objective is to design a gain matrix K such that the system (6.5) is quadratically stable.

The result of designing the sliding surface can be stated as the following theorem.

Theorem 6.1 *The reduced-order system* (6.5) *is quadratically stable if there exist symmetric positive definite matrices* $Z = Z^T$, $Z_d \in \mathbb{R}^{(n-m)\times(n-m)}$ *and a general matrix* $Y \in \mathbb{R}^{m\times(n-m)}$ *such that the following LMI holds:*

$$\begin{bmatrix} Z - Z_d & * & * \\ 0 & Z_d - Z - Z^T & * \\ \bar{A}_{11}Z - \bar{A}_{12}KZ & \bar{A}_{d11}Z - \bar{A}_{d12}KZ & -Z \end{bmatrix} < 0. \tag{6.6}$$

Moreover, the gain matrix K is given by

$$K = YZ^{-1}. \tag{6.7}$$

Proof Choose a Lyapunov–Krasovsky candidate functional

$$V(\xi_1(k), k) = \xi_1^T(k)Z^{-1}\xi_1(k) + \sum_{\alpha=k-k_x}^{k-1} \xi_1^T(\alpha)Z_d^{-1}\xi_1(\alpha), \tag{6.8}$$

which is positive definite $\forall \xi_1(k) \neq 0$. The first-order difference equation of the Lyapunov function is

$$V(k+1) = \xi_1^T(k+1)Z^{-1}\xi_1(k+1) + \sum_{\alpha=k-k_x+1}^{k} \xi_1^T(\alpha)Z_d^{-1}\xi_1(\alpha); \qquad (6.9)$$

the inequality $\Delta V(k) = V(k+1) - V(k) < 0$ is known to show that the trajectory of the system can be driven to the sliding mode plane in finite time. Then

$$
\begin{aligned}
\Delta V(k) &= V(k+1) - V(k) \\
&= \xi_1^T(k)[\tilde{A}_{11}^T Z^{-1}\tilde{A}_{11} + Z_d^{-1} - Z^{-1}]\xi_1(k) + \xi_1^T(k)\tilde{A}_{11}^T Z^{-1}\tilde{A}_{d11}\xi_1(k-k_x) \\
&\quad + \xi_1^T(k-k_x)\tilde{A}_{d11}^T Z^{-1}\tilde{A}_{11}\xi_1(k) \\
&\quad + \xi_1^T(k-k_x)[\tilde{A}_{d11}^T Z^{-1}\tilde{A}_{d11} - Z_d^{-1}]\xi_1(k-k_x),
\end{aligned}
\qquad (6.10)
$$

where $\tilde{A}_{11} = (\bar{A}_{11} - \bar{A}_{12}K))$ and $\tilde{A}_{d11} = (\bar{A}_{d11} - \bar{A}_{d12}K)$ for simplicity.

If

$$\begin{bmatrix} \tilde{A}_{11}^T Z^{-1}\tilde{A}_{11} + Z_d^{-1} - Z^{-1} & \tilde{A}_{11}^T Z^{-1}\tilde{A}_{d11} \\ \tilde{A}_{d11}^T Z^{-1}\tilde{A}_{11} & \tilde{A}_{d11}^T Z^{-1}\tilde{A}_{d11} - Z_d^{-1}, \end{bmatrix} < 0 \qquad (6.11)$$

then $\Delta V(k) < 0$, and $\left[\xi_1^T(k) \quad \xi_1^T(k-k_x)\right]^T \neq 0$, which is equivalent to

$$\begin{bmatrix} Z_d^{-1} - Z^{-1} & 0 \\ 0 & -Z_d^{-1} \end{bmatrix} + \begin{bmatrix} \tilde{A}_{11}^T \\ \tilde{A}_{d11}^T \end{bmatrix} Z^{-1} \begin{bmatrix} \tilde{A}_{11} & \tilde{A}_{d11} \end{bmatrix} < 0. \qquad (6.12)$$

Using the Schur complement [29], which is equivalent to

$$\begin{bmatrix} Z_d^{-1} - Z^{-1} & * & * \\ 0 & -Z_d^{-1} & * \\ \tilde{A}_{11} & \tilde{A}_{d11} & -Z \end{bmatrix} < 0, \qquad (6.13)$$

pre- and postmultiplying both sides of the LMI by $diag\{Z, Z, I_{n-m}\}$ makes the above inequality equivalent to

$$\begin{bmatrix} -Z + ZZ_d^{-1}Z & * & * \\ 0 & -ZZ_d^{-1}Z & * \\ \bar{A}_{11}Z - \bar{A}_{12}Y & \bar{A}_{d11}Z - \bar{A}_{d12}Y & -Z \end{bmatrix} < 0. \qquad (6.14)$$

The following inequality is introduced to separate Z from Z_d^{-1} [30, 31]. It is known that

$$(Z - Z_d)Z_d^{-1}(Z - Z_d) = ZZ_d^{-1}Z - Z - Z^T + Z_d \geq 0,$$

which implies that

$$-ZZ_d^{-1}Z \leq Z_d - Z - Z^T.$$

Thus we can obtain (6.6). It is seen that the inequality $\Delta V(k) < 0$ holds $\forall \, (\xi_1(k), k) \in \mathbb{R}^{n-m}$. Therefore, the reduced-order system (6.5) is quadratically stable with $K = YZ^{-1}$.

Moreover, the quadratically stable sliding surface (6.4) is

$$s(k) = YZ^{-1}\xi_1(k) + \xi_2(k) = 0.$$

Hence the proof is complete.

Remark 6.1 It is easy to see that inequality (6.6) of Theorem 6.1 can be cast as an LMI with variable Z, Z_d, and Y. When implemented with the given matrices, this set of LMIs turns out to be feasible, and every solution defines a gain matrix $K = YZ^{-1}$ that solves the problem. LMI facilitates the design of controllers in the convex optimisation approach, which is computationally superior to other approaches and easily solvable using available LMI solver algorithms [32].

6.3.2 Discrete-Time Sliding Mode Control Design

Our objective is to find the SMC that achieves the same objective as in (1.18) for the system (6.1). It can be proved using the following theorem.

> **Theorem 6.2** *There exist computable values of $W_c, m_1^{\pm}(k), m_2^{\pm}(k)$ such that the time-delay system with DSMC law*
>
> 1. $u(k) = u^+(k) \in (cB)^{-1}(-cAx(k) - cA_d x(k - k_x)) + [m_1^+(k), m_2^+(k)])$ *if* $s(k) > \theta^{-1}W_c$,
> 2. $u(k) = u^-(k) \in (cB)^{-1}(-cAx(k) - cA_d x(k - k_x)) + [m_1^-(k), m_2^-(k)])$ *if* $s(k) < -\theta^{-1}W_c$,
> 3. $u(k) = -(cB)^{-1}[cAx(k) + cA_d x(k - k_x)]$ *if* $|s(k)| \le \theta^{-1}W_c$,
>
> *drives the system states (6.1) to the SMB $S_{\mu c}$ and holds it within that band with the given probability $(1 - \delta)$, where W_c is as defined in Chap. 1.*

Proof Three cases are given for the following control law.

Case 6.1 $s(k) > \theta^{-1}W_c$
　　On considering (1.19) and (1.28), the SMC law is obtained to be

$$u(k) = u^+(k) \in (cB)^{-1}(-c(Ax(k) + A_d x(k - k_x)) + [m_1^+(k), m_2^+(k)]) \quad (6.15)$$

where $m_i^+(k), i = 1, 2$, is defined in Chap. 1.

Case 6.2 $s(k) < -\theta^{-1}W_c$.

Similarly, on considering (1.19) and (1.34), the control strategy will be

$$u(k) = u^-(k) \in (cB)^{-1}(-c(Ax(k) + A_d x(k - k_x)) + [m_1^-(k), m_2^-(k)]), \quad (6.16)$$

where $m_i^-(k)$, $i = 1, 2$, is as defined in Chap. 1.

Case 6.3 $|s(k)| \le \theta^{-1}W_c$.

In this case, the equivalent SMC should be adopted as $m_1 = m_2 = 0$:

$$u(k) = -(cB)^{-1}c(Ax(k) + A_d x(k - k_x)). \quad (6.17)$$

From Lemma 1.1, it can be seen that $u(k) = -(cB)^{-1}c(Ax(k) + A_d x(k - k_x))$ will be within the valid range of values for $u(k)$. Hence it can be used for the SMC implementation. It is also valid for Case 1 and 2.

Remark 6.2 When the DSMC (6.17) is used, it will be

$$s(k + 1) = c\Gamma w(k). \quad (6.18)$$

Then the system's motion (1.15) is called SSM.

The procedures of the proposed DSMC for stochastic systems are summarised as follows:

Algorithm 6.1 DSMC Design for Delayed Stochastic Systems

1: Calculate the gain matrix K in equation (6.7) by solving the LMI equation (6.6) in Theorem 6.1.
2: Calculate the sliding function matrix c such that Assumption 6.4 is satisfied.
3: Construct the sliding mode control (6.17).

6.4 Design of DSMC for Stochastic Systems with Incomplete State Information

In this section, incomplete states are estimated by the Kalman filter approach for discrete-time delayed stochastic systems. State estimation for the linear discrete-time delayed stochastic system procedure is described in Algorithm 6.2.

Algorithm 6.2 State Estimation for Time Delayed Discrete-time System

1: Measurement update "Correction" equations

$$a) \Upsilon(k) = P(k|k-1)C^T[CP(k|k-1)C^T + GRG^T]^{-1}.$$
$$b) \hat{x}(k|k) = \hat{x}(k|k-1) + \Upsilon(k)[y(k) - C\hat{x}(k|k-1)] \qquad (6.19)$$
$$c) P(k|k) = [I - \Upsilon(k)C]P(k|k-1)$$

where $\Upsilon(k)$ is the Kalman gain and estimation error is describe as $\tilde{x}(k) \triangleq x(k) - \hat{x}(k)$. Given the co-variance $P(k)$ and $P(k-k_x)$ of the error vector $\tilde{x}(k)$ and $\tilde{x}(k-k_x)$, respectively.

2: Time update "Prediction" equations

$$a) \hat{x}(k|k-1) = A\hat{x}(k-1|k-1) + A_d\hat{x}(k-k_x-1|k-k_x-1) + Bu(k-1),$$
$$b) P(k|k-1) = AP(k-1|k-1)A^T + A_dP(k-k_x-1|k-k_x-1)A_d^T \qquad (6.20)$$
$$+ A_dP(k-k_x-1|k-k_x-1)A^T + \Gamma Q\Gamma^T$$

Remark 6.3 The philosophy behind Algorithm 6.2 is that the computational aspects of control should be divided into two parts. First, the state vector should be constructed; this is the main job of the Kalman filter as an observer. Second, the design the SMC should be implemented.

6.4.1 Controller Design Strategy

Theorem 6.3 *There exist computable values of* $\tilde{W}(k+1)$, $\tilde{m}_1^{\pm}(k)$, $\tilde{m}_2^{\pm}(k)$ *such that the estimated state time-delay system of the SMC law*

1. $u(k) = u^+(k) \in (cB)^{-1}(-cA\hat{x}(k) - cA_d\hat{x}(k-k_x) + [\tilde{m}_1^+(k), \tilde{m}_2^+(k)])$ *if* $s(k) > \theta^{-1}\tilde{W}(k+1)$,
2. $u(k) = u^-(k) \in (cB)^{-1}(-cA\hat{x}(k) - cA_d\hat{x}(k-k_x) + [\tilde{m}_1^-(k), \tilde{m}_2^-(k)])$ *if* $s(k) < -\theta^{-1}\tilde{W}(k+1)$,
3. $u(k) = -(cB)^{-1}[cA\hat{x}(k) + cA_d\hat{x}(k-k_x)]$ *if* $|s(k)| \leq \theta^{-1}\tilde{W}(k+1)$,

drives the system's states (6.1) and (6.2) to the SMB $S_{\mu c}$ *and held within that band with probability* $(1-\delta)$, *where the solution of* $\tilde{W}(k+1)$ *is as defined in Chap. 1.*

Proof The control laws are described as follows:

Case 6.4 If $\hat{s}(k) > \theta^{-1}\tilde{W}(k+1)$, then the SMC law is

$$u(k) = u^+(k) \in (cB)^{-1}(-c(A\hat{x}(k) + A_d\hat{x}(k-k_x)) + [\tilde{m}_1^+(k), \tilde{m}_2^+(k)]),$$
$$(6.21)$$

where $\tilde{m}_i^+(k)$, $i = 1, 2$, is as defined in Chap. 1.

Case 6.5 If $\hat{s}(k) < -\theta^{-1}\tilde{W}(k+1)$, then the SMC law is

$$u(k) = u^-(k) \in (cB)^{-1}(-c(A\hat{x}(k) + A_d\hat{x}(k - k_x)) + [\tilde{m}_1^-(k), \tilde{m}_2^-(k)]),$$

(6.22)

where $\tilde{m}_i^-(k)$, $i = 1, 2$, is as defined in Chap. 1.

Case 6.6 If $|\hat{s}(k)| \leq \theta^{-1}\tilde{W}(k+1)$, then the control should be adopted as $m_1 = m_2 = 0$:

$$u(k) = -(cB)^{-1}c(A\hat{x}(k) + A_d\hat{x}(k - k_x)).$$

(6.23)

It can be noted that $u(k) = -(cB)^{-1}c(A\hat{x}(k) + A_d\hat{x}(k - k_x))$ will be within the valid range of values for $u(k)$. Hence it can be used for the SMC implementation for estimated states. This controller is valid for the all three of the above-mentioned cases.

Remark 6.4 When the equivalent DSMC (6.23) is used, the sliding variable $s(k)$ will have the dynamics

$$\hat{s}(k+1) \triangleq c\Upsilon(k+1)[CA\tilde{x}(k) + CA_d\tilde{x}(k - k_x) + C\Gamma w(k) + Gv(k+1)],$$

(6.24)

and the system's motion is referred to as the stochastic sliding mode in the time-delayed system.

6.4.2 Simulation Results

In this section, we give an example of a General Electric F404 aircraft engine to illustrate the effectiveness of the obtained result. The matrices of the system (6.1) are [13]

$$A = \begin{bmatrix} 0.2504 & 0 & 0.3919 \\ 0.0570 & 0.6188 & -0.0616 \\ 0.0502 & 0 & 0.1262 \end{bmatrix}, A_d = \begin{bmatrix} 0.03 & 0 & -0.01 \\ 0.02 & 0.03 & 0 \\ 0.04 & 0.05 & -0.01 \end{bmatrix},$$

$$B = \begin{bmatrix} 0.1857 & 0.4286 \\ 0.1597 & 0.793 \\ 0.1138 & 0.0581 \end{bmatrix}, \Gamma = \begin{bmatrix} 0.1857 \\ 0.1597 \\ 0.1138 \end{bmatrix}, C = \begin{bmatrix} 1 & 0 & 0 \\ 0 & 1 & 0 \end{bmatrix}, G = \begin{bmatrix} 1 \\ 0 \end{bmatrix}.$$

Following the design procedures in the complete state information, the SMC is given by the following steps.

Step 1: The LMI (6.6) is solved using the LMI toolbox in MATLAB [33]. Based on Theorem 6.1, the solutions are obtained at a given time as follows:

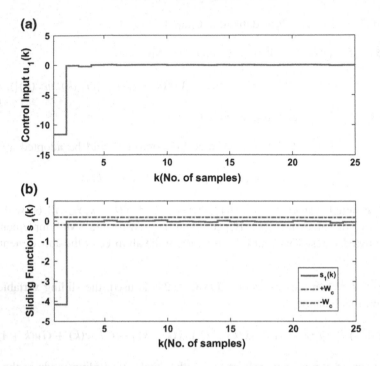

Fig. 6.1 Complete state information: **a** control input $u_1(k)$ and **b** sliding function $s_1(k)$

$$Z = 59.5649, \, Z_d = 79.3365, \, Y = \begin{bmatrix} -7.3538 \,\, 0.3752 \end{bmatrix}^T.$$

Then the sliding gain matrix from (6.7) can be obtained as

$$K = \begin{bmatrix} -0.1235 \\ 0.0063 \end{bmatrix}.$$

Step 2: The sliding function is computed as follows:

$$s(k) = \begin{bmatrix} -0.4107 & -0.9037 & -0.1731 \\ 0.5380 & -0.3871 & 0.7488 \end{bmatrix} x(k).$$

Step 3: The control input is obtained as

$$u(k) = \begin{bmatrix} -1.1827 & 1.6598 & -2.5279 \\ 0.1342 & -1.0613 & 0.5321 \end{bmatrix} x(k)$$

$$+ \begin{bmatrix} -0.3007 & -0.2054 & 0.0991 \\ 0.0387 & 0.0151 & -0.0199 \end{bmatrix} x(k - k_x).$$

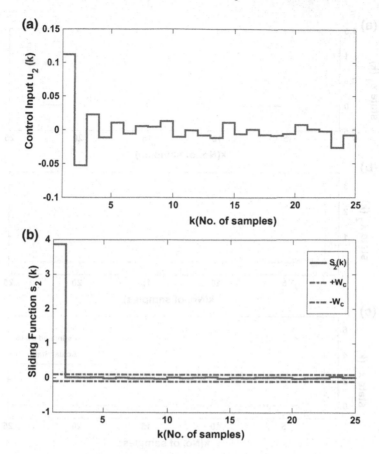

Fig. 6.2 Complete state information: **a** control input $u_2(k)$ and **b** sliding function $s_2(k)$

The covariance matrices for input and output noise sequences are given as $Q = 0.01I_3$ and $R = 0.01I_2$. The primary values of the system states and delayed state are determined by $x(0) = [1 \quad 3 \quad 6]^T$ and $x(-1) = [1 \quad 2 \quad 5]^T$. The primary values of the estimated state and delayed state are $\hat{x}(0) = [5 \quad 6 \quad 4]^T$ and $\hat{x}(-1) = [2 \quad 4 \quad 3]^T$. The error covariances of the state and delayed states are $P(0) = 0.1I_2$ and $P(-1) = 0.01I_2$, respectively. The parameters θ and ε are chosen as $\theta = 0.9$ and $\varepsilon = 0.05$. Figure 6.3 shows the time responses of different state trajectories. It can be clearly seen that the system states approach zero. Figures 6.1 and 6.2 show the sliding functions and control inputs for complete state information. The sliding surface variables converge to a neighbourhood of zero, which verifies that the proposed DSMC ensures the existence of a sliding mode. Since the proposed method does not need a switching type of control law, it can be seen that no chattering phenomenon will occur. The obtained results clearly show that the suggested method provides good system performance for a discrete time-delay system and simultaneously ensures

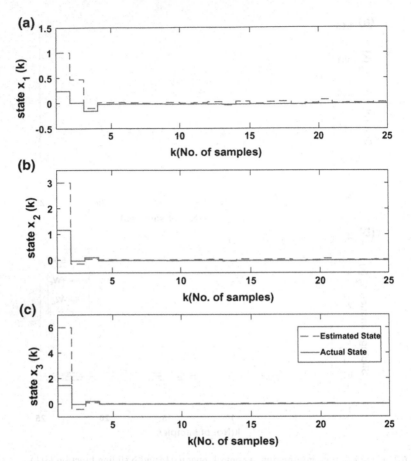

Fig. 6.3 Response for **a** state $x_1(k)$ and its state estimation $\hat{x}_1(k)$; **b** state $x_2(k)$ and its state estimation $\hat{x}_2(k)$; **c** state $x_3(k)$ and its state estimation $\hat{x}_3(k)$

stability and robustness against state delay and noise. Figures 6.4 and 6.5 show the estimated control inputs and sliding functions for incomplete state information. After the state estimation, sliding functions are also obtained in the sliding surface. It is seen that the state trajectory converges to the SMB. These simulation results demonstrate that our proposed design is very effective.

In the next section, we propose a functional observer-based SMC for a discrete-time delayed stochastic system.

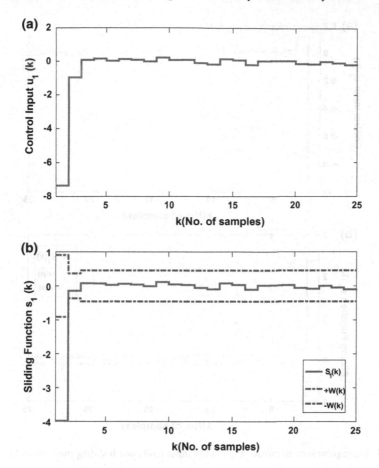

Fig. 6.4 Incomplete state information: **a** control input $u_1(k)$ and **b** sliding function $s_1(k)$

6.5 Functional Observer-Based SMC Design for Time Delayed Stochastic Systems

Our results cover two aspects in this section: the general case and the case of an internal delay-free observer.

6.5.1 General Case

In this section, our goal is to design a functional state observer in order to estimate the control $u_f = Lx(k) + L_d x(k - k_x)$ for a given system (6.1). The matrices L

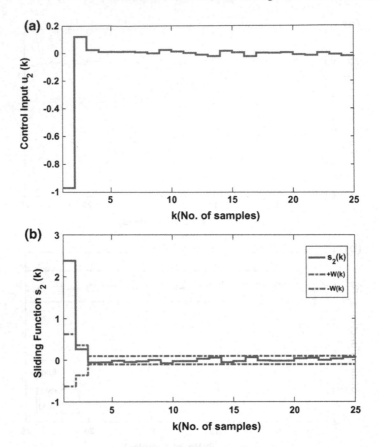

Fig. 6.5 Incomplete state information: **a** control input $u_2(k)$ and **b** sliding function $s_2(k)$

and L_d are obtained by the control input (6.17), where $L = -(cB)^{-1}CA$ and $L_d = -(cB)^{-1}CA_d$.

Our goal is to design an observer of the form

$$\zeta(k+1) = M\zeta(k) + M_d\zeta(k-k_x) + Jy(k) + J_dy(k-k_x) + Hu(k)$$
$$\hat{u}_f(k) = V\zeta(k) + V_d\zeta(k-k_x) + Ey(k) + E_dy(k-k_x),$$
(6.25)

where $\zeta(k) \in \mathbb{R}^q$ is the state vector, and $\hat{u}_f(k) \in \mathbb{R}^r$ is desired functional estimate of $u(k)$. The matrices $M, M_d \in \mathbb{R}^{q \times q}$, $J, J_d \in \mathbb{R}^{q \times p}$, $V, V_d \in \mathbb{R}^{r \times q}$, $E, E_d \in \mathbb{R}^{r \times p}$, and $H \in \mathbb{R}^{q \times m}$ are unknown constants.

Theorem 6.4 *The qth-order observer (6.25) will estimate $u_f(k)$ if the following conditions hold:*

(i) M and M_d are stable matrices;
(ii) $MT + JC - TA = 0$;
(iii) $M_d T + J_d C - T A_d = 0$;
(iv) $VT = L - EC$;
(v) $H = TB$;
(vi) $V_d T = L_d - E_d C$;
(vii) $q \geq \rho(L(I_n - C^+ C))$,

where $L, L_d \in \mathbb{R}^{r \times n}$ are the functional gain matrices and $T \in \mathbb{R}^{q \times n}$ is an unknown constant matrix.

Proof Let us define $e(k)$ as the error between $\zeta(k)$ and $Tx(k)$ as

$$e(k) \triangleq \zeta(k) - Tx(k). \tag{6.26}$$

Taking the first-order difference equation of (6.26) gives

$$
\begin{aligned}
e(k+1) &= \zeta(k+1) - Tx(k+1) \\
&= M\zeta(k) + M_d\zeta(k - k_x) + Jy(k) + J_d y(k - k_x) + Hu(k) - TAx(k) \\
&\quad - TA_d x(k - k_x) - TBu(k) - T\Gamma w(k) \\
&= Me(k) + M_d e(k - k_x) + (MT + JC - TA)x(k) + (H - TB)u(k) \\
&\quad + (M_d T + J_d C - TA_d)x(k - k_x) + JGv(k) + J_d Gv(k - k_x) - T\Gamma w(k).
\end{aligned}
\tag{6.27}
$$

The independence of the input error dynamics (6.27) from $x(k), x(k - k_x), u(k)$ requires that

$$MT + JC - TA = 0, \tag{6.28}$$

$$M_d T + J_d C - TA_d = 0, \tag{6.29}$$

and

$$H - TB = 0. \tag{6.30}$$

Then the error dynamics (6.27) becomes

$$e(k+1) = Me(k) + M_d e(k - k_x) + JGv(k) + J_d Gv(k - k_x) - T\Gamma w(k). \tag{6.31}$$

The estimation error of the output is represented as

$$e_u(k) \triangleq \hat{u}_f(k) - Lx(k) - L_d x(k - k_x)$$
$$= V\zeta(k) + Ey(k) + E_d y(k - k_x) - Lx(k) - L_d x(k - k_x)$$
$$= Ve(k) + (VT + EC - L)x(k) + E_d Gv(k - k_x) + (E_d C - L_d)x(k - k_x)$$
$$+ EGv(k).$$

(6.32)

The independence of the output error dynamics from $x(k)$ and $x(k - k_x)$ requires that

$$VT = L - EC \tag{6.33}$$

and

$$V_d T = L_d - E_d C. \tag{6.34}$$

Then the output estimation error becomes

$$e_u(k) = Ve(k) + EGv(k) + E_d Gv(k - k_x). \tag{6.35}$$

Hence the output error dynamics $e_u(k)$ is governed by the matrices V, V_d, E, and E_d. The matrices E and E_d are given as $E = LC^+$ and $E_d = L_d C^+$ to satisfy the observer condition (vii) of Theorem 6.4. Further, the unknown terms J, J_d, and T in the observer Eq. (6.25) can be solved as follows.

Let

$$\begin{bmatrix} J & J_d & T \end{bmatrix} = \mathscr{X}, \tag{6.36}$$

where $\mathscr{X} \in \mathbb{R}^{q \times (n+2p)}$ is an unknown matrix.

Using (6.36), the matrices J, J_d, and T can be expressed in terms of the unknown matrix $\mathscr{X}$ as

$$J = \mathscr{X} \begin{bmatrix} I \\ 0 \\ 0 \end{bmatrix} \tag{6.37}$$

$$J_d = \mathscr{X} \begin{bmatrix} 0 \\ I \\ 0 \end{bmatrix} \tag{6.38}$$

and

$$T = \mathscr{X} \begin{bmatrix} 0 \\ 0 \\ I \end{bmatrix}. \tag{6.39}$$

On putting the values of J and T in (6.28),

$$M\mathscr{X} \begin{bmatrix} 0 \\ 0 \\ I \end{bmatrix} + \mathscr{X} \begin{bmatrix} C \\ 0 \\ -A \end{bmatrix} = 0. \tag{6.40}$$

On putting the values of J_d and T in (6.29), one obtains

$$
M_d \mathscr{X} \begin{bmatrix} 0 \\ 0 \\ I \end{bmatrix} + \mathscr{X} \begin{bmatrix} 0 \\ C \\ -A_d \end{bmatrix} = 0 \tag{6.41}
$$

$$
V \mathscr{X} \begin{bmatrix} 0 \\ 0 \\ I \end{bmatrix} = L - EC. \tag{6.42}
$$

Now assembling the matrices (6.40)–(6.42) in composite form yields the following:

$$
\begin{bmatrix} M \\ M_d \\ V \end{bmatrix} \mathscr{X} \begin{bmatrix} 0 \\ 0 \\ I \end{bmatrix} + \begin{bmatrix} I \\ 0 \\ 0 \end{bmatrix} \mathscr{X} \begin{bmatrix} C \\ 0 \\ -A \end{bmatrix} + \begin{bmatrix} 0 \\ I \\ 0 \end{bmatrix} \mathscr{X} \begin{bmatrix} 0 \\ C \\ -A_d \end{bmatrix} = \Theta, \tag{6.43}
$$

with

$$
\Theta = \begin{bmatrix} 0 \\ 0 \\ L - EC \end{bmatrix},
$$

where $\Theta \in \mathbb{R}^{(2q+r)\times n}$ is a known constant matrix.

By applying the Kronecker product [34] in (6.43), we obtain

$$
\Sigma vec(\mathscr{X}) = vec(\Theta), \tag{6.44}
$$

with

$$
\Sigma = \begin{bmatrix} \begin{bmatrix} 0 \\ 0 \\ I \end{bmatrix}^T \otimes \begin{bmatrix} M \\ M_d \\ V \end{bmatrix} + \begin{bmatrix} C \\ 0 \\ -A \end{bmatrix}^T \otimes \begin{bmatrix} I \\ 0 \\ 0 \end{bmatrix} + \begin{bmatrix} 0 \\ C \\ -A_d \end{bmatrix}^T \otimes \begin{bmatrix} 0 \\ I \\ 0 \end{bmatrix} \end{bmatrix},
$$

where $vec(\mathscr{X}) \in \mathbb{R}^{(2p+n)q}$, $vec(\Theta) \in \mathbb{R}^{n(2q+r)}$, and $\Sigma \in \mathbb{R}^{(2q+r)n \times q(n+2p)}$.

The error covariance of (6.31) propagates as

$$
\begin{aligned}
P(k+1) &= MP(k)M^T + M_d P(k-k_x)M_d^T + M_d P(k-k_x, k)M^T + JGR(k)G^T J^T \\
&\quad + J_d GR(k-k_x)G^T J_d^T + T\Gamma Q\Gamma^T T^T.
\end{aligned} \tag{6.45}
$$

The output error covariance of (6.35) is

$$
P_u(k) = VP(k)V^T + EGRG^T E^T + E_d GR_{(k-k_x)}G^T E_d^T. \tag{6.46}
$$

We can now write $P_u(k+1)$ in terms of the covariance as

$$
P_u(k+1) = VP(k+1)V^T + EGR_k G^T E^T + E_d GR_{(k-k_x)}G^T E_d^T. \tag{6.47}
$$

To minimise the effect of process and measurement noise covariance, we choose $\mathscr{X}$ in (6.36) as

$$\mathscr{X} = \underset{\mathscr{X}}{\arg\min} \left\| V \mathscr{X} \begin{bmatrix} GR_k G^T & 0 & 0 \\ 0 & GR_{k-k_x} G^T & 0 \\ 0 & 0 & \Gamma Q \Gamma^T \end{bmatrix} \mathscr{X}^T V^T \right\|$$

such that the effect of noise in the functional observer is minimised.

Designing a q-dimensional dynamical system where

$$q \geq \rho(L(I_n - C^+ C))$$

completes the proof.

Based on the aforementioned progress, the design procedure is now summarised as follows:

Algorithm 6.3 Functional Observer based SMC for Time Delayed Stochastic Systems

---

1: To obtain the sliding function $s(k) \triangleq cx(k)$
2: Design the stable SMC $u(k)$, and functional to be estimated is $Lx(k) + L_d x(k - k_x)$, where $L = -(cB)^{-1} cA$ and $L_d = -(cB)^{-1} cA_d$, respectively.
3: To obtain the order of the functional observer, $q \geq \rho(L(I_n - C^+ C))$, such that $rank(V) = rank(L - EC)$.
4: Choose the $(r \times q)$ elements of matrix V, arbitrarily.
5: Choose, matrices M and M_d are stable matrices.
6: From (6.37), (6.38) and (6.39), find the values of matrix J, J_d and T.
7: Now, find H by $H = TB$.
8: Solve composite form of matrix (6.43) by Kronecker product, for matrix $\mathscr{X}$. Matrix $\mathscr{X}$ is chosen according to

$$\mathscr{X} = \underset{\mathscr{X}}{\arg\min} \left\| \mathscr{X} \begin{bmatrix} GR_k G^T & 0 & 0 \\ 0 & GR_{k-k_x} G^T & 0 \\ 0 & 0 & \Gamma Q \Gamma^T \end{bmatrix} \mathscr{X}^T \right\|$$

subject to the equality condition (6.43), if satisfied, go to next step; else set $q = q + 1$ and go to step 3.
9: As an outcome of steps 1–8 above, recover a structure of functional state observer by (6.25).

6.5.2 Functional Observer Without Internal Delay

In this subsection, we present the implementation of the previous subsection's results to the special case in which (6.25) is independent of $\zeta(k - k_x)$, in which case the observer is called an observer without internal delay [21]. If $M_d = 0$ and $V_d = 0$, the structure of the observer (6.25) becomes a functional observer without internal delay of the form

$$\zeta(k+1) = M\zeta(k) + Jy(k) + J_d y(k - k_x) + Hu(k),$$
$$\hat{u}_f(k) = V\zeta(k) + Ey(k) + E_d y(k - k_x),$$

(6.48)

and the conditions of Theorem 6.4 become

  (i) M is stable matrix;
 (ii) $MT + JC - TA = 0$;
(iii) $J_d C - TA_d = 0$;
 (vi) $VT = L - EC$;
  (v) $H = TB$;
 (vi) $L_d - E_d C = 0$;
(vii) $q \geq \rho(L(I_n - C^+ C))$.

Equations (6.31) and (6.45) are reduced to

$$e(k+1) = Me(k) + JGv(k) + J_d Gv(k - k_x) - T\Gamma w(k)$$

(6.49)

and

$$P(k+1) = MP(k)M^T + JGR(k)G^T J^T + J_d GR(k - k_x)G^T J_d^T + T\Gamma Q\Gamma^T T^T.$$

(6.50)

6.6 Simulation Example and Results

In this section, simulation results will be given to validate the advantage and effectiveness of the approaches proposed [24]:

$$A = \begin{bmatrix} 0.3 & -0.1 & 0.1 & -0.2 \\ 0.6 & -0.2 & 0.2 & -0.4 \\ 0.1 & 0.3 & -0.2 & 0 \\ -0.1 & 0 & -0.1 & -0.3 \end{bmatrix}, A_d = \begin{bmatrix} 0.2 & 0.1 & 0 & -0.1 \\ -0.2 & -0.1 & 0 & 0.1 \\ 0.1 & 0.2 & -0.1 & 0 \\ 0.2 & 0.1 & 0.3 & -0.1 \end{bmatrix}, B = \Gamma = \begin{bmatrix} 1 \\ 2 \\ 0 \\ 1 \end{bmatrix}.$$

The parameters θ and ε are chosen as $\theta = 0.9$ and $\varepsilon = 0.05$ [35]. The state delay is $k_x = 1$. The initial conditions of the state and delayed state are selected, arbitrarily, as $x(0) = [1 \ 3 \ 6 \ 4]^T$, $x_d(-1) = [-1 \ 2 \ 5 \ 1]^T$. The initial conditions of the observer state and delayed state observer are chosen, arbitrarily, as $\zeta(0) = [1 \ 2]^T$, $\zeta_d(-1) = [2 \ 3]^T$;

$$C = \begin{bmatrix} I_3 & 0_{3\times1} \end{bmatrix}, G = \begin{bmatrix} 1 & 0 & 0 \end{bmatrix}^T.$$

6.6.1 Comparative Study

Apart from the functional observer-based state feedback and SMC law, the other parameter is kept the same in order to make a fair comparison.

6.6.1.1 Functional Observer-Based SMC

The LMI (6.6) is solved using the LMI toolbox of MATLAB [33]. Based on Theorem 6.1, all the solutions are obtained as follows:

$$Z = \begin{bmatrix} 24.345 & 1.305 & -0.262 \\ 1.305 & 24.315 & 0.574 \\ -0.262 & 0.574 & 25.544 \end{bmatrix}, Z_d = \begin{bmatrix} 32.969 & 1.305 & -0.262 \\ 1.305 & 32.939 & 0.574 \\ -0.262 & 0.574 & 34.168 \end{bmatrix},$$

and

$$Y = \begin{bmatrix} -6.215 & -3.481 & -17.838 \end{bmatrix}.$$

Then the sliding gain matrix from (6.7) can be obtained as

$$K = \begin{bmatrix} -0.256 & -0.112 & -0.698 \end{bmatrix}.$$

The sliding function is obtained as

$$s(k) = \begin{bmatrix} 0.903 & 0.846 & -0.112 & -0.146 \end{bmatrix} x(k).$$

The control input is chosen to be

$$u(k) = \begin{bmatrix} -0.378 & 0.119 & -0.121 & 0.194 \end{bmatrix} x(k)$$
$$+ \begin{bmatrix} 0.012 & 0.012 & 0.013 & -0.003 \end{bmatrix} x(k - k_x).$$

Now the matrices E and E_d are obtained as

$$E = \begin{bmatrix} -0.319 & 0.119 & -0.121 \end{bmatrix}, E_d = \begin{bmatrix} 0.0120 & 0.0129 & 0.0134 \end{bmatrix}.$$

Here the order of the observer is $q = 2$.

A $(q \times q)$ diagonal convergent matrix is chosen as

$$M = \begin{bmatrix} 0.1 & 0 \\ 0 & 0.4 \end{bmatrix}, M_d = \begin{bmatrix} 0.3 & 0 \\ 0 & 0.2 \end{bmatrix}.$$

Consider arbitrary matrices V and V_d as follows:

$$V = \begin{bmatrix} 1 & 2 \end{bmatrix}, V_d = \begin{bmatrix} 3 & 4 \end{bmatrix}.$$

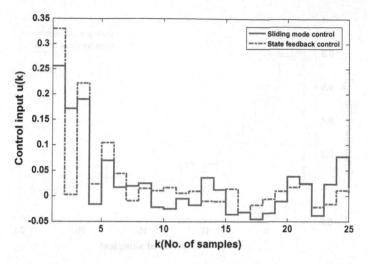

Fig. 6.6 Evolution of control input $u(k)$

Now the matrix $\mathscr{X}$ is obtained as $\mathscr{X} = [\mathscr{X}_1 \quad \mathscr{X}_2]$, where

$$\mathscr{X}_1 = \begin{bmatrix} -0.124 & 0.048 & -0.099 & 0.130 & -0.131 \\ 0.095 & -0.973 & 0.171 & 0.009 & 0.330 \end{bmatrix}$$

and

$$\mathscr{X}_2 = \begin{bmatrix} -0.093 & 0.002 & 0.042 & 0.002 & 0.066 \\ -0.021 & 0.485 & -0.085 & -0.005 & -0.073 \end{bmatrix}.$$

Solving (6.37) and (6.38) gives the following matrices J and J_d:

$$J = \begin{bmatrix} -0.124 & 0.048 & -0.099 \\ -0.093 & 0.002 & 0.042 \end{bmatrix}, \quad J_d = \begin{bmatrix} 0.130 & -0.131 & 0.095 \\ 0.002 & 0.066 & -0.021 \end{bmatrix},$$

and solving (6.39) for the matrix T gives

$$T = \begin{bmatrix} -0.973 & 0.171 & 0.009 & 0.330 \\ 0.485 & -0.089 & -0.005 & -0.073 \end{bmatrix}.$$

Since $H = TB$, the matrix H is obtained as

$$H = \begin{bmatrix} -0.301 \\ 0.241 \end{bmatrix}.$$

Thus the estimate of $u(k)$ is given by the functional observer (6.25).

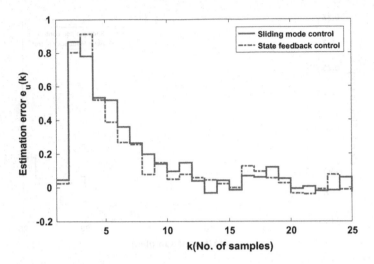

Fig. 6.7 Evolution of estimation error $e_u(k)$

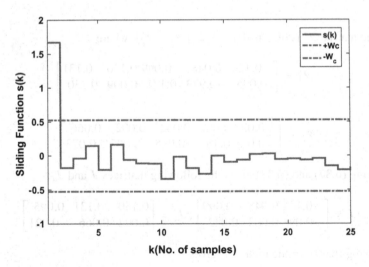

Fig. 6.8 Evolution of sliding function $s(k)$

6.6.1.2 Functional Observer-Based State Feedback

The functional to be estimated is $Lx(k) + L_d x(k - k_x)$, where L and L_d are obtained by the state feedback method. The remaining parameter is obtained using the approach as mentioned above and can be chosen in the same way:

$$u(k) = \begin{bmatrix} -0.22 & 0.08 & -0.06 & 0.22 \end{bmatrix} x(k) + \begin{bmatrix} 0.04 & 0.02 & -0.06 & -0.02 \end{bmatrix} x(k - k_x).$$

Table 6.1 Performance comparison of different control strategies

Functional observer-based control algorithm	$\sum_k |u(k)|$	$\sum_k |e_u(k)|$
State feedback control	1.2	31.548
Sliding mode control	0.745	28.429

The parameters M, M_d, and V are same as in the previous case. Finally, we obtained the structure of the functional observer as

$$E = \begin{bmatrix} -0.22 & 0.08 & -0.06 \end{bmatrix}, \, E_d = \begin{bmatrix} 0.04 & 0.02 & -0.06 \end{bmatrix}$$

$$J = \begin{bmatrix} -0.1406 & 0.0554 & -0.1122 \\ -0.1057 & 0.0018 & 0.0471 \end{bmatrix}, \, J_d = \begin{bmatrix} 0.1475 & -0.1483 & 0.1079 \\ 0.0024 & 0.0748 & -0.0231 \end{bmatrix}$$

$$H = \begin{bmatrix} 0.7622 \\ -0.2771 \end{bmatrix}.$$

Thus the estimate of $u(k)$ is given by the functional observer (6.25).

The simulation results are presented in Figs. 6.6, 6.7, 6.8, 6.9, 6.10 and 6.11. On comparing the proposed functional observer-based DSMC and functional observer-based state feedback control, the proposed DSMC shows the superiority in disturbance reduction. The DSMC settles within a much smaller magnitude than that of the SFC method in Fig. 6.6. The estimation error of DSMC and SFC is presented in Fig. 6.7. The value of $u(k)$ remains around 0.004 in the steady state with SFC.

On the other hand, in the DSMC algorithm, the value of $u(k)$ is around 0.002, which is closer to zero than the steady-state value obtained using SFC. Hence the estimation error settles within a smaller magnitude than that of the SFC method. Thus the proposed algorithm shows a better result.

To compare the proposed algorithm with SFC numerically, the integral absolute estimation error (IAEE) $\sum_k |e_u(k)|$ and integral absolute control input (IACI) $\sum_k |u(k)|$ are used. In Table 6.1, it is observed that the proposed LFO-based SMC gives a smaller integral absolute estimation error of 28.43. This results in a smaller control effort compared to LFO-based SFC. It also proves that the lesser IAEE does not come at the cost of a larger control input. Hence the proposed algorithm can achieve a much better performance than the SFC-based algorithm in the considered example.

Figure 6.8 shows the sliding function response. The sliding function $s(k)$ lies within the band, which is defined by $W_c = 0.5465$.

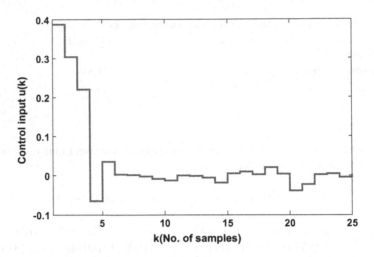

Fig. 6.9 Evolution of control input $u(k)$ without internal delay

6.6.2 Functional Observer-Based SMC Without Internal Delay

Using a similar operation, the structure of the functional observer form without internal delay can be expressed as

$$M = \begin{bmatrix} 0.1 & 0 \\ 0 & 0.4 \end{bmatrix}, J = \begin{bmatrix} -0.263 & 0.096 & -0.125 \\ 0.035 & -0.073 & 0.047 \end{bmatrix}, J_d = \begin{bmatrix} 0.004 & -0.004 & 0.139 \\ 0.017 & 0.011 & -0.041 \end{bmatrix},$$

$$H = \begin{bmatrix} -0.452 \\ 0.323 \end{bmatrix}, V = \begin{bmatrix} 1 & 2 \end{bmatrix}, E = \begin{bmatrix} -0.319 & 0.119 & -0.122 \end{bmatrix}, E_d = \begin{bmatrix} 0.012 & 0.013 & 0.014 \end{bmatrix}.$$

Thus the estimate of $u(k)$ is given by the functional observer (6.48).

Figures 6.9, 6.10 and 6.11 show respectively the response of the control input $u(k)$, the estimated error $e_u(k)$, and the sliding function response $s(k)$ without internal delay. The sliding function $s(k)$ lies within the band defined by $W_c = 0.523$. These simulation results demonstrate that the proposed design is very adequate.

6.7 Conclusions

In this chapter, the SMC design problem for discrete-time delayed stochastic systems has been considered. A tactic has been successfully used to achieve the sliding mode in discrete-time stochastic systems for complete state information. In the complete state information case, the sliding function lies within a specified band. Further, the stochastic SMC is designed for a system with incomplete state information. It is noted

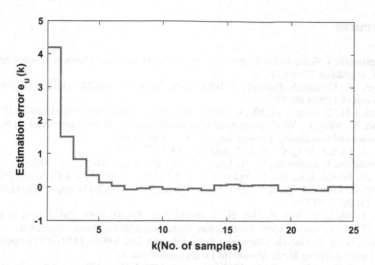

Fig. 6.10 Evolution of estimation error $e_u(k)$ without internal delay

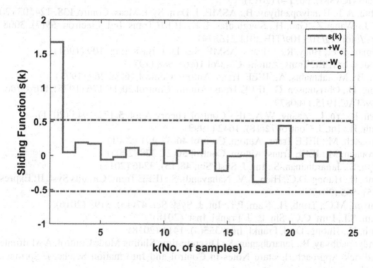

Fig. 6.11 Evolution of the sliding function $s(k)$ without internal delay

that a sliding function lies within a specified band in the case of incomplete state information. Furthermore, a functional observer-based based SMC has been designed for a discrete-time delayed stochastic system. The obtained theoretical results have been validated by given simulation results.

References

1. Kharitonov, V.: Time-Delay Systems: Lyapunov Functionals and Matrices. Control Engineering. Birkhäuser, Boston (2012)
2. Cheres, E., Gutman, S., Palmor, Z.J.: IEEE Trans. Autom. Control **34**(11), 1199 (1989). https://doi.org/10.1109/9.40753
3. Choi, H.H.: Electron. Lett. **30**(13), 1100 (1994). https://doi.org/10.1049/el:19940732
4. Liao, X., Wang, L., Yu, P.: Stability of Dynamical Systems. Monograph Series on Nonlinear Science and Complexity. Elsevier Science, New York (2007)
5. Xu, S., Lam, J., Yang, C.: Syst. Control Lett. **43**(2), 77 (2001)
6. Gouaisbaut, F., Dambrine, M., Richard, J.: Syst. Control Lett. **46**(4), 219 (2002)
7. Shi, P., Boukas, E.K., Shi, Y., Agarwal, R.K.: J. Comput. Appl. Math. **157**(2), 435 (2003)
8. Song, H., Chen, S.C., Yam, Y.: IEEE Trans. Cybern. **PP**(99), 1 (2017). https://doi.org/10.1109/TCYB.2016.2577340
9. Hu, J., Wang, Z., Niu, Y., Gao, H.: J. Frankl. Inst. **351**(4), 2185 (2014). Special Issue on 2010-2012 Advances in Variable Structure Systems and Sliding Mode Algorithms
10. Hu, J., Wang, Z., Gao, H., Stergioulas, L.K.: J. Frankl. Inst. **349**(4), 1459 (2012). Special Issue on Optimal Sliding Mode Algorithms for Dynamic Systems
11. Singh, S., Janardhanan, S.: In: 2015 International Workshop on Recent Advances in Sliding Modes (RASM), pp. 1–6 (2015)
12. Mehta, A.J., Bandyopadhyay, B.: ASME, J. Dyn. Syst. Meas. Contro **138**, 124503 (2016)
13. Hu, J., Wang, Z., Gao, H., Stergioulas, L.K.: IEEE Trans. Ind. Electron. **59**(7), 3008 (2012). https://doi.org/10.1109/TIE.2011.2168791
14. Kalman, R.E., Bucy, R.S.: Trans. ASME. Ser. D. J. Basic Eng. 109 (1961)
15. Rhodes, I.: IEEE Trans. Autom. Control **16**(6), 688 (1971)
16. Rao, B., Mahalanabis, A.: IEEE Trans. Autom. Control **16**(3), 267 (1971)
17. Liang, D., Christensen, G.: IEEE Trans. Autom. Control **20**(1), 176 (1975). https://doi.org/10.1109/TAC.1975.1100879
18. Chen, B., Yu, L., Zhang, W.A.: IET Control Theory Appl. **5**(17), 1945 (2011)
19. Trinh, H.: Int. J. Control **72**(18), 1642 (1999)
20. Darouach, M.: IEEE Trans. Autom. Control **46**(3), 491 (2001)
21. Darouach, M.: IEEE Trans. Autom. Control **50**(2), 228 (2005)
22. Singh, S., Janardhanan, S.: Int. J. Syst. Sci. **48**(15), 3246 (2017)
23. Trinh, H., Huong, D.C., Hien, L.V., Nahavandi, S.: IEEE Trans. Circuits Syst. II: Express Briefs **64**(5), 555 (2017)
24. Nguyen, M.C., Trinh, H., Nam, P.T.: Int. J. Syst. Sci. **47**(13), 3193 (2016)
25. Islam, S.I., Lim, C.C., Shi, P.: J. Frankl. Inst. (2018)
26. Trinh, H., Huong, D.: J. Frankl. Inst. **355**(3), 1411 (2018)
27. Bandyopadhyay, B., Janardhanan, S.: Discrete-time Sliding Mode Control: A Multirate Output Feedback Approach. Lecture Notes in Control and Information Sciences. Springer, Berlin (2005)
28. Koshkouei, A.J., Zinober, A.S.I.: ASME **122**(4), 793 (2000). https://doi.org/10.1115/1.1321266
29. Xia, Y., Jia, Y.: IEEE Trans. Autom. Control **48**(6), 1086 (2003)
30. Xia, Y., Fu, M., Shi, P., Wang, M.: IET Control Theory Appl. **4**(4), 613 (2010)
31. Xia, Y., Liu, G.P., Shi, P., Chen, J., Rees, D.: Int. J. Robust Nonlinear Control **18**(11), 1142 (2008)
32. Scherer, C., Weiland, S.: The Control Systems Handbook. Linear Matrix Inequalities in Control, 2nd edn. (2010)
33. Gahinet, P.: LMI Control Toolbox: For Use with MATLAB ; User's Guide ; Version 1. Computation, visualization, programming (MathWorks) (1995)
34. Brewer, J.: IEEE Trans. Circuits Syst. **25**(9), 772 (1978)
35. Sage, A., Melsa, J.: Estimation Theory with Applications to Communications and Control. McGraw-Hill Series in Systems Science. McGraw-Hill, New York (1971)

Chapter 7
Functional Observer-Based Sliding Mode Control for Parametric Uncertain Discrete-Time Delayed Stochastic Systems

Abstract This chapter is concerned with the problem of the functional observer-based sliding mode control (SMC) design for parametric uncertain discrete-time delayed stochastic systems including mismatched parameter uncertainty in the state matrix and in the delayed state matrix. Stability analysis of the sliding function is presented in a time delayed stochastic system with the linear matrix inequality (LMI) approach. Moreover, it is shown that the state trajectories can be driven onto the specified sliding surface despite the presence of state delay, unmatched parameter uncertainty, and stochastic noise in the system. The research is motivated by the fact that system states are not always accessible for the state feedback. Therefore, SMC is estimated using the functional observer technique. To mitigate the side effect of parametric uncertainty on the estimation error, a sufficient condition of stability is proposed based on Gershgorin's circle theorem. The claims made are validated through numerical simulations.

Keywords Discrete-time systems · Stochastic system · State delay · Parametric uncertainty · Sliding mode control · State estimation · Linear matrix inequalities · Linear functional observer

7.1 Introduction

As mentioned in Chap. 6, it is shown that a functional observer-based SMC may perform better than the state feedback control method in certain cases. Although the results presented in Chap. 6 are effective, it is not straightforward to apply these algorithms when the system matrices (A, A_d) contain parametric uncertainties. Most of the studies available in the literature do not tackle the presence of parametric uncertainty in time delayed systems [1, 2]. However, parametric uncertainties are always present in the state and delayed state matrix, which is not discussed in Chap. 6. Designing a functional observer without considering the parametric uncertainty may lead to imperfect estimation and deterioration of closed-loop performance [3, 4], which means that the estimation error should converge asymptotically to zero in the absence of disturbances [5–11]. In other words, when the system matrices are affected

© Springer Nature Switzerland AG 2020

S. Singh and S. Janardhanan, *Discrete-Time Stochastic Sliding Mode Control Using Functional Observation*, Lecture Notes in Control and Information Sciences 483, https://doi.org/10.1007/978-3-030-32800-9_7

by parametric uncertainties, the functional observer must minimise the action of the system states on the estimated error [3]. Despite the importance of the estimation of the system states under certain parameters due to modelling errors, to the best of the authors' knowledge, there is no work in which the problem of designing functional observer-based SMC for time-delayed stochastic systems is investigated. With this motivation, this chapter seeks to explore the use of functional observer-based SMC for time-delayed stochastic systems in the case of parametric uncertainty present in the state and delayed state matrix. In sliding function design, system uncertainties are considered to enhance the system's robustness.

The contributions of this chapter lie in the following three aspects [12]:

1. A sliding function is designed for parametric uncertain time delayed stochastic systems with uncertainties in each of the system matrices.
2. The SMC is calculated using a functional observer approach in the presence of parametric uncertainty in the state and delayed state matrix.
3. To mitigate the side effect of the parametric uncertainty on the estimation error in the time delayed state, a sufficient condition on stability is proposed based on Gershgorin's circle theorem.

The rest of the chapter is organised as follows. Section 7.2 introduces the SMC design problem. Section 7.3 presents the sliding function and SMC design for uncertain discrete-time delayed stochastic systems. A functional observer-based SMC design analysis for discrete-time delayed systems is presented in Sect. 7.4. The effectiveness and applicability of the proposed method is illustrated through a numerical example in Sect. 7.5. Finally, in Sect. 7.6, some remarks conclude the chapter.

7.2 Problem Formulation

Consider a discrete-time parametric uncertain delayed stochastic system:

$$
\begin{aligned}
x(k + 1) &= (A + \Delta A)x(k) + (A_d + \Delta A_d)x(k - k_x) + Bu(k) + \Gamma w(k), \\
x(k) &= \phi(k), \forall k \in [-k_x, \quad 0], \\
y(k) &= Cx(k) + Gv(k),
\end{aligned}
\tag{7.1}
$$

where $x(k)$, $x(k - k_x)$, $u(k)$, and $y(k)$ are matrices of appropriate dimensions. In addition, $k_x > 0$ is the known constant state delay. The terms $\Delta A \in \mathbb{R}^{n \times n}$ and $\Delta A_d \in \mathbb{R}^{n \times n}$ do not satisfy the matching conditions. Let the initial state x_0 and x_{-k_x} be random vectors with zero mean and given covariance matrices P_0 and P_{-k_x}, respectively. The following assumptions are useful for the development of the proposed technique.

Assumption 7.1 The system (7.1) is controllable and observable.

Assumption 7.2 The initial state $x(0)$ and delayed state $x(-kx)$ are mutually correlated Gaussian random vectors, and $x(0), x(-kx), w(k),$ and $v(k)$ are mutually uncorrelated.

Assumption 7.3 The system parametric uncertainties are norm-bounded of the following form:

$$[\Delta A \quad \Delta A_d] = G_1 \Delta(k)[E_1 \quad E_2].$$

Here $G_1 \in \mathbb{R}^{n \times n_a}$, $E_1 \in \mathbb{R}^{n_b \times n}$ and $E_2 \in \mathbb{R}^{n_b \times n}$ are known real constant matrices, and the properly dimensioned matrix $\Delta(k) \in \mathbb{R}^{n_a \times n_b}$ is unknown but norm-bounded as $\Delta^T(k)\Delta(k) \leq I$, $\forall k \geq 0$.

Remark 7.1 The parametric uncertainties ΔA and ΔA_d represent the rarity of an exact mathematical model of a dynamic system due to the system's complexity. The uncertainty has been used extensively in many practical systems that can be either exactly modelled or bounded above by the condition $\Delta^T(k)\Delta(k) \leq I$. The matrix $\Delta(k)$ contains the uncertain parameters, and the constant matrices G_1, E_1, and E_2 specify how the uncertain parameter $\Delta(k)$ disturbs the nominal matrix of system (7.1).

The main aim of this chapter is to design a functional observer-based SMC for a parametric uncertain time delayed system (7.1) with norm-bounded uncertainties on each of the system matrices such that the sliding mode is achieved.

7.3 Design of Sliding Function and Controller

As defined in (1.9), the sliding function is the same as for the uncertain time delayed system (7.1).

7.3.1 Design of Sliding Function

For the purpose of designing the sliding function, the system (7.1) is transformed into regular form. Consider the matrix U as defined in (5.2).

Let $\xi(k) = Ux(k)$. Then

$$\begin{aligned}
\xi_1(k+1) &= (\bar{A}_{11} + \Delta\bar{A}_{11})\xi_1(k) + (\bar{A}_{12} + \Delta\bar{A}_{12})\xi_2(k) + (\bar{A}_{d11} + \Delta\bar{A}_{d11})\xi_1(k - kx) \\
&\quad + (\bar{A}_{d12} + \Delta\bar{A}_{d12})\xi_2(k - kx) \\
\xi_2(k+1) &= (\bar{A}_{21} + \Delta\bar{A}_{21})\xi_1(k) + (\bar{A}_{22} + \Delta\bar{A}_{d22})\xi_2(k) + (\bar{A}_{d21} + \Delta\bar{A}_{d21})\xi_1(k - kx) \\
&\quad + (\bar{A}_{d22} + \Delta\bar{A}_{d22})\xi_2(k - kx) + B_2 u(k) + \Gamma_2 w(k),
\end{aligned}$$

$$(7.2)$$

where

$$U(A + \Delta A)U^{-1} = \begin{bmatrix} \bar{A}_{11} + \Delta\bar{A}_{11} & \bar{A}_{12} + \Delta\bar{A}_{12} \\ \bar{A}_{21} + \Delta\bar{A}_{21} & \bar{A}_{22} + \Delta\bar{A}_{22} \end{bmatrix},$$

$$U(A_d + \Delta A_d)U^{-1} = \begin{bmatrix} \bar{A}_{d11} + \Delta\bar{A}_{d11} & \bar{A}_{d12} + \Delta\bar{A}_{d12} \\ \bar{A}_{d21} + \Delta\bar{A}_{d21} & \bar{A}_{d22} + \Delta\bar{A}_{d22} \end{bmatrix},$$

and

$$U\Gamma = \begin{bmatrix} 0 \\ \Gamma_2 \end{bmatrix}.$$

Then the sliding function (1.9) is expressed in terms of the new state $\xi(k)$ as

$$s(k) \triangleq cU^T\xi(k) \triangleq K\xi_1(k) + \xi_2(k), \tag{7.3}$$

where the gain is $K \in \mathbb{R}^{m \times (n-m)}$. Substituting $\xi_2(k) = -K\xi_1(k)$ in the first equation of system (7.2) gives

$$\begin{aligned} \xi_1(k+1) &= (\bar{A}_{11} + \Delta\bar{A}_{11} - \bar{A}_{12}K - \Delta\bar{A}_{12}K)\xi_1(k) \\ &+ (\bar{A}_{d11} + \Delta\bar{A}_{d11} - \bar{A}_{d12}K - \Delta\bar{A}_{d12}K)\xi_1(k - k_x). \end{aligned} \tag{7.4}$$

The objective is to design a gain matrix K such that the system (7.4) is quadratically stable.

Assumption 7.4 cB is a nonsingular matrix.

Remark 7.2 The sliding function has been defined in (7.3), and designing a sliding function is equivalent to obtaining the constant matrix K. But it is difficult to determine the gain K for the robust stability of the system (7.4), because the gain matrix K occurs not only in the system matrix $\bar{A}_{11}$ but also in the delayed system matrix $\bar{A}_{d11}$ [10].

The essential results of sliding function design are recapitulated in Theorem 7.1.

Theorem 7.1 *The reduced-order uncertain system (7.4) is quadratically stable if there exist symmetric positive definite matrices $Z = Z^T$, $Z_d \in \mathbb{R}^{(n-m) \times (n-m)}$, a matrix $Y \in \mathbb{R}^{m \times (n-m)}$, and a scalar $\lambda > 0$ such that the following LMI holds:*

$$\begin{bmatrix} Z - Z_d & * & * & * \\ 0 & Z_d - Z - Z^T & * & * \\ E_{11}Z - E_{12}KZ & E_{d11}Z - E_{d12}KZ & -\lambda I & * \\ \bar{A}_{11}Z - \bar{A}_{12}KZ & \bar{A}_{d11}Z - \bar{A}_{d12}KZ & 0 & -Z + \lambda G_1 G_1^T \end{bmatrix} < 0. \tag{7.5}$$

Moreover, the gain matrix K is given by

$$K = YZ^{-1}. \tag{7.6}$$

Proof. Consider a Lyapunov–Krasovsky functional as follows:

$$V(k) = \xi_1^T(k)Z^{-1}\xi_1(k) + \sum_{\alpha=k-k_x}^{k-1} \xi_1^T(\alpha)Z_d^{-1}\xi_1(\alpha). \tag{7.7}$$

It is positive if any $\forall \xi_1(k) \neq 0$ for $i \in (k - k_x, k)$. The first-order difference equation of the Lyapunov function is

$$V(k+1) = \xi_1^T(k+1)Z^{-1}\xi_1(k+1) + \sum_{\alpha=k-k_x+1}^{k} \xi_1^T(\alpha)Z_d^{-1}\xi_1(\alpha). \tag{7.8}$$

The inequality $\Delta V(k) = V(k+1) - V(k) < 0$ shows that the trajectory of the system can be stably driven when confined to the sliding function

$$
\begin{aligned}
\Delta V(k) &= V(k+1) - V(k) \\
&= \xi_1^T(k)[\tilde{A}_{11}^T Z^{-1}\tilde{A}_{11} + Z_d^{-1} - Z^{-1}]\xi_1(k) + \xi_1^T(k)\tilde{A}_{11}^T Z^{-1}\tilde{A}_{d11}\xi_1(k-k_x) \\
&\quad + \xi_1^T(k-k_x)\tilde{A}_{d11}^T Z^{-1}\tilde{A}_{11}\xi_1(k) \\
&\quad + \xi_1^T(k-k_x)[\tilde{A}_{d11}^T Z^{-1}\tilde{A}_{d11} - Z_d^{-1}]\xi_1(k-k_x),
\end{aligned}
\tag{7.9}
$$

where $\tilde{A}_{11} = (\bar{A}_{11} + \Delta\bar{A}_{11} - \bar{A}_{12}K - \Delta\bar{A}_{12}K)$, $\tilde{A}_{d11} = (\bar{A}_{d11} + \Delta\bar{A}_{d11} - \bar{A}_{d12}K - \Delta\bar{A}_{d12}K)$.

If

$$
\begin{bmatrix}
\tilde{A}_{11}^T Z^{-1}\tilde{A}_{11} + Z_d^{-1} - Z^{-1} & \tilde{A}_{11}^T Z^{-1}\tilde{A}_{d11} \\
\tilde{A}_{d11}^T Z^{-1}\tilde{A}_{11} & \tilde{A}_{d11}^T Z^{-1}\tilde{A}_{d11} - Z_d^{-1}
\end{bmatrix} < 0,
\tag{7.10}
$$

then $\Delta V(k) < 0$ for $\left[\xi_1^T(k) \; \xi_1^T(k-k_x)\right]^T \neq 0$.

Equation (7.10) is equivalent to

$$
\begin{bmatrix}
Z_d^{-1} - Z^{-1} & 0 \\
0 & -Z_d^{-1}
\end{bmatrix}
+
\begin{bmatrix}
\tilde{A}_{11}^T \\
\tilde{A}_{d11}^T
\end{bmatrix}
Z^{-1}
\begin{bmatrix} \tilde{A}_{11} & \tilde{A}_{d11} \end{bmatrix} < 0.
\tag{7.11}
$$

By Lemma 5.2, Eq. (7.11) can be shown to be equivalent to

$$
\begin{bmatrix}
Z_d^{-1} - Z^{-1} & * & * \\
0 & -Z_d^{-1} & * \\
\tilde{A}_{11} & \tilde{A}_{d11} & -Z
\end{bmatrix} < 0.
\tag{7.12}
$$

The inequality (7.12) can be further written as

$$
\begin{bmatrix}
Z_d^{-1} - Z^{-1} & * & * \\
0 & -Z_d^{-1} & * \\
(\bar{A}_{11} - \bar{A} A_{12} K) & (\bar{A}_{11} - \bar{A}_{12} K) & -Z
\end{bmatrix}
+ \begin{bmatrix} 0 \\ 0 \\ G_1 \end{bmatrix} \Delta(k) \left[(E_{11} - E_{12}K)\ (E_{d11} - E_{d12}K)\ 0 \right]
$$

$$
+ \begin{bmatrix}
(E_{11} - E_{12}K)^T \\
(E_{d11} - E_{d12}K)^T \\
0
\end{bmatrix} \Delta^T(k) \left[0\ 0\ G_1^T \right] < 0.
$$

$$(7.13)$$

In light of Lemma 5.1, the inequality (7.13) holds for all $\Delta^T(k)$ satisfying $\Delta^T(k)\Delta(k) \le I$ if and only if there exists a scalar $\lambda > 0$ such that

$$
\begin{bmatrix}
Z_d^{-1} - Z^{-1} & * & * \\
0 & -Z_d^{-1} & * \\
(\bar{A}_{11} - \bar{A}_{12}K) & (\bar{A}_{d11} - \bar{A}_{d12}K) & -Z
\end{bmatrix}
+ \lambda \begin{bmatrix} 0 \\ 0 \\ G_1 \end{bmatrix} \left[0\ 0\ G_1^T \right]
$$

$$
+ \lambda^{-1} \begin{bmatrix}
(E_{11} - E_{12}K)^T \\
(E_{d11} - E_{d12}K)^T \\
0
\end{bmatrix} \left[(E_{11} - E_{12}K)\ (E_{d11} - E_{d12}K)\ 0 \right] < 0.
$$

$$(7.14)$$

Using the Schur complement [10], Eq. (7.14) can be shown to be equivalent to

$$
\begin{bmatrix}
Z_d^{-1} - Z^{-1} & * & * & * \\
0 & -Z_d^{-1} & * & * \\
(E_{11} - E_{12}K) & (E_{d11} - E_{d12}K) & -\lambda I & * \\
(\bar{A}_{11} - \bar{A}_{12}K) & (\bar{A}_{d11} - \bar{A}_{d12}K) & 0 & -Z + \lambda G_1 G_1^T
\end{bmatrix} < 0.
$$

$$(7.15)$$

Pre- and postmultiplying inequality (7.15) by the matrix $\text{diag}\{Z, Z, I_{n-m}, I_{n-m}\}$ yields

$$
\begin{bmatrix}
-Z + Z Z_d^{-1} Z & * & * & * \\
0 & -Z Z_d^{-1} Z & * & * \\
E_{11}Z - E_{12}KZ & E_{d11}Z - E_{d12}KZ & -\lambda I & * \\
\bar{A}_{11}Z - \bar{A}_{12}KZ & \bar{A}_{d11}Z - \bar{A}_{d12}KZ & 0 & -Z + \lambda G_1 G_1^T
\end{bmatrix} < 0.
$$

$$(7.16)$$

From the equality [2, 9]

$$
(Z - Z_d) Z_d^{-1} (Z - Z_d) = Z Z_d^{-1} Z - Z - Z^T + Z_d \geqslant 0 \quad \forall Z, Z_d
$$

we conclude that

$$
Z Z_d^{-1} Z \geqslant Z + Z^T - Z_d,
$$

$$(7.17)$$

and we can write Eq. (7.16) as

$$
\begin{bmatrix}
Z - Z_d & * & * & * \\
0 & Z_d - Z - Z^T & * & * \\
E_{11}Z - E_{12}Y & E_{d11}Z - E_{d12}Y & -\lambda I & * \\
\bar{A}_{11}Z - \bar{A}_{12}Y & \bar{A}_{d11}Z - \bar{A}_{d12}Y & 0 & -Z + \lambda G_1 G_1^T
\end{bmatrix} < 0, \tag{7.18}
$$

where $Y = KZ$. It is seen that the above inequality (7.18) is $\Delta V(k) < 0 \, \forall \, (\xi_1(k), k) \in \mathbb{R}^{(n-m)}$. Therefore, reduced-order systems is quadratically stable with $K = YZ^{-1}$.

Moreover, the quadratically stable sliding function of (7.3) is

$$
s(k) = YZ^{-1}\xi_1(k) + \xi_2(k) = 0,
$$

which completes the proof.

Remark 7.3 If $Z_d = \eta Z$, then the LMI (7.18) can be reduced to

$$
\begin{bmatrix}
Z(1 - \eta) & * & * & * \\
0 & -\eta Z & * & * \\
E_{11}Z - E_{12}Y & E_{d11}Z - E_{d12}Y & -\lambda I & * \\
\bar{A}_{11}Z - \bar{A}_{12}Y & \bar{A}_{d11}Z - \bar{A}_{d12}Y & 0 & -Z + \lambda G_1 G_1^T
\end{bmatrix} < 0,
$$

with $\eta > 1$.

Remark 7.4 A special case obtains if the perturbation ΔA and ΔA_d are matched. In this case, the LMI (7.18) reduces to [2]

$$
\begin{bmatrix}
Z - Z_d & * & * \\
0 & Z_d - Z - Z^T & * \\
\bar{A}_{11}Z - \bar{A}_{12}Y & \bar{A}_{d11}Z - \bar{A}_{d12}Y & -Z
\end{bmatrix} < 0.
$$

Remark 7.5 A second possible case of system (7.1) occurs when there are no time delays, i.e., $A_d = 0$ and $\Delta A_d = 0$. The condition (7.18) in Theorem 7.1 then simplifies to [13]

$$
\begin{bmatrix}
Z & * \\
\bar{A}_{11}Z - \bar{A}_{12}Y & -Z
\end{bmatrix} < 0.
$$

7.3.2 Design of DSMC

Our main goal is to find the DSMC that achieves the same objective as in (1.18) for the system (7.1). The disturbance bounds are known [14]:

$$
d_l \leq d_m(k) = c(\Delta A x(k) + \Delta A_d x(k - k_x)) \leq d_u,
$$

where the lower bound d_l and the upper bound d_u are known constants. Furthermore, we introduce the following notation:

$$d_0 = (d_l + d_u)/2,$$

where the average value of $d_m(k)$ is denoted by d_0.

The rest of the analysis of the SMC is the same as is Chap. 6. The SMC is obtained as

$$u(k) = -(cB)^{-1}(cAx(k) + cA_dx(k - k_x) + d_0), \qquad (7.19)$$

and the variance is given by $\sigma_c^2 = c\Gamma Q\Gamma^T c^T$.

Remark 7.6 When the DSMC (7.19) is used, the sliding variable $s(k)$ will have the dynamics

$$s(k + 1) = c\Gamma w(k) + d_m(k) - d_0.$$

Then the system's motion (7.1) is called the stochastic sliding mode.

In the next section, we introduce functional observer-based SMC for discrete-time delayed stochastic systems.

7.4   Design of Functional Observer-Based SMC for Parametric Uncertain Time-Delay Systems

7.4.1   General Case

Our intention is to design a functional state observer of the form (6.25), the same as in Chap. 6. Similar to the general case mentioned Sect. 6.5, the estimate of $u_f(k) = Lx(k) + L_dx(k - k_x)$, using the qth-order observer given (6.25) for the system (7.1), can be obtained, provided the condition stated in Theorem 6.4 holds. Due to the presence of parametric uncertainty in the system dynamics (7.1), the error dynamics of the error $e(k) \triangleq \zeta(k) - Tx(k)$ will be different from (6.31), and it can be obtained as

$$
\begin{aligned}
e(k + 1) = {} & Me(k) + M_de(k - k_x) + (MT + JC - TA - T\Delta A)x(k) + (H - TB)u(k) \\
& + (M_dT + J_dC - TA_d - T\Delta A_d)x(k - k_x) + JGv(k) + J_dGv(k - k_x) \\
& - T\Gamma w(k).
\end{aligned}
$$

$$(7.20)$$

On the other hand, from (6.25), the estimate of $u(k)$ can be expressed as

$$
\begin{aligned}
\hat{u}_f(k) = {} & Ve(k) + V_de(k - k_x) + (VT + EC)x(k) + (V_dT + E_dC)x(k - k_x) \\
& + E_dGv(k - k_x) + EGv(k).
\end{aligned}
$$

$$(7.21)$$

If conditions (ii)–(vi) of Theorem 6.4 are satisfied, then (7.20) and (7.21) reduce to

$$e(k+1) = Me(k) + M_d e(k-k_x) - T\Delta Ax(k) - T\Delta A_d x(k-k_x) + JGv(k)$$
$$+ J_d Gv(k-k_x) - T\Gamma w(k)$$

$$(7.22)$$

and

$$e_u(k) \triangleq \hat{u}_f(k) - Lx(k) - L_d x(k-k_x)$$
$$= Ve(k) + V_d e(k-k_x) + EGv(k) + E_d Gv(k-k_x),$$

$$(7.23)$$

where $E = LC^+$ and $E_d = L_d C^+$. On the other hand, to mitigate the side effect of the parametric uncertainties on the dynamics of the estimation error (7.22), a sufficient condition for stability is given by Gershgorin's circle theorem.

The composite form of (7.1) and (7.22) is as follows:

$$\mathcal{G}(k+1) = (\mathcal{K} + \Delta\mathcal{K})\mathcal{G}(k) + (\mathcal{K}_d + \Delta\mathcal{K}_d)\mathcal{G}(k-k_x) + \mathcal{B}w(k) + \mathcal{B}_1 v(k)$$
$$+ \mathcal{B}_d v(k-k_x),$$

$$(7.24)$$

where $\mathcal{G}(k) = [x^T(k)\ \ e^T(k)]^T$, $\mathcal{K} = \begin{bmatrix} A+BL & 0 \\ 0 & M \end{bmatrix}$,

$$\Delta\mathcal{K} = \begin{bmatrix} \Delta A & 0 \\ -T\Delta A & 0 \end{bmatrix} = \begin{bmatrix} G_1 \\ -TG_1 \end{bmatrix} \Delta(k)[E_1\ 0], \quad \mathcal{K}_d = \begin{bmatrix} A_d+BL_d & 0 \\ 0 & M_d \end{bmatrix},$$

$$\Delta\mathcal{K}_d = \begin{bmatrix} \Delta A_d & 0 \\ -T\Delta A_d & 0 \end{bmatrix} = \begin{bmatrix} G_1 \\ -TG_1 \end{bmatrix} \Delta(k)[E_2\ 0], \quad \mathcal{B} = \begin{bmatrix} \Gamma \\ -T\Gamma \end{bmatrix},$$

$$\mathcal{B}_1 = \begin{bmatrix} 0 \\ JG \end{bmatrix}, \quad \mathcal{B}_d = \begin{bmatrix} 0 \\ J_d G \end{bmatrix}.$$

Using Gershgorin's circle theorem, we deduce that if each eigenvalue of $\mathcal{K}$ lies in the union of the circles, then the system will be stable. In that case, the following condition holds: $|\mathscr{Z} - (\mathcal{K}_{ii} + \Delta\mathcal{K}_{ii}) - (\mathcal{K}_d ii + \Delta\mathcal{K}_d)z^{-k_x})| \le r_i(\mathcal{K}_{ii} + \mathcal{K}d_{ii}z^{-k_x})$.

The values of J, J_d, and T are obtained using the similar method described in Sect. 6.5 of Chap. 6.

The error covariance of (7.24) can be reproduced as

$$P(k+1) = \mathcal{K}P(k)\mathcal{K}^T + \mathcal{K}_d P(k-k_x)\mathcal{K}_d^T + \mathcal{K}_d P(k-k_x,k)\mathcal{K}^T + \mathcal{B}Q\mathcal{B}^T$$
$$+ \mathcal{B}_1 R(k)\mathcal{B}_1^T + \mathcal{B}_d R(k-k_x)\mathcal{B}_d^T.$$

$$(7.25)$$

To minimise the effect of the process and measurement noise covariance, we choose $\mathscr{X}$ as

$$\mathscr{X} = \underset{\mathscr{X}}{\text{argmin}} \left\| \mathscr{X} \begin{bmatrix} 0 & 0 \\ 0 & \begin{bmatrix} GR_k G^T & 0 & 0 \\ 0 & GR_{k-k_x}G^T & 0 \\ 0 & 0 & \Gamma Q\Gamma^T \end{bmatrix} \end{bmatrix} \mathscr{X}^T \right\|$$

$$(7.26)$$

in order to minimise the effect of noise in the functional observer.

The algorithm for designing a functional observer-based SMC law can be outlined as follows:

Algorithm 7.1 Summary of Linear Functional Observer-Based SMC for Time-Delayed Stochastic Systems

1: To solve the LMI problem (7.5) and obtain K to construct the sliding gain c as in (7.3), and then find the values of $L = -(cB)^{-1}cA$ and $L_d = -(cB)^{-1}cA_d$.

2: Choose the matrices M, M_d, V, and V_d arbitrarily.

3: To acquire the order of the LFO, $q \geq \rho(L(I_n - C^+C))$, such that $rank(V) = rank(L - EC)$,

4: The matrix $\mathscr{X}$ is chosen according to (7.26) and subject to the equality condition (7.26). If this is done successfully, go to the next step; otherwise, set $q = q + 1$ and go to step 3.

5: Find the values of the matrices J, J_d, and T.

6: Now find H by $H = TB$.

7: As a result of steps 1–6 above, a similar structure of the functional observer to that in (6.25), the DSMC (7.19), and the sliding function (7.3) are obtained.

7.4.2 Internal Delay-Free Observer

We present an application of the above results to the particular case in which (6.25) is independent of $\zeta(k - k_x)$ [15]. If $M_d = 0$ and $V_d = 0$, the structure of the observer (6.25) becomes a functional observer without internal delay of the form (6.48), and the conditions of Theorem 6.4 become the same as in Sect. 6.5.2 of Chap. 6.

Equations (7.22) and (7.25) are reduced to

$$e(k + 1) = Me(k) - T\Delta Ax(k) - T\Delta A_d x(k - k_x) + JGv(k) + J_d Gv(k - k_x) - T\Gamma w(k) \tag{7.27}$$

and

$$\mathscr{G}(k + 1) = (\mathscr{K} + \Delta\mathscr{K})\mathscr{G}(k) + (\mathscr{K}_d + \Delta\mathscr{K}_d)\mathscr{G}(k - k_x) + \mathscr{B}w(k) + \mathscr{B}_1 v(k) + \mathscr{B}_d v(k - k_x), \tag{7.28}$$

where $\mathscr{G}(k) = [x^T(k) \quad e^T(k)]^T$, $\mathscr{K} = \begin{bmatrix} A + BL + \Delta A & 0 \\ -T\Delta A & M \end{bmatrix}$,

$\mathscr{K}_d = \begin{bmatrix} A_d + BL_d + \Delta A_d & 0 \\ -T\Delta A_d & 0 \end{bmatrix}$, $\mathscr{B} = \begin{bmatrix} \Gamma \\ -T\Gamma \end{bmatrix}$, $\mathscr{B}_1 = \begin{bmatrix} 0 \\ JG \end{bmatrix}$, $\mathscr{B}_d = \begin{bmatrix} 0 \\ J_d G \end{bmatrix}$.

The rest of the analysis will be same as that for Theorem 6.4.

7.5 Simulation Example and Results

A description of the numerical example dynamics is given in Sect. 6.6 of Chap. 6. The parametric uncertainties in the state matrix are $\Delta A = G_1 \Delta(k) E_1$, and those in the delayed state matrix are $\Delta A_d = G_1 \Delta(k) E_2$ with $\Delta(k) = 0.5\sin(k)$,

$$G_1 = [0.4 \ -0.2 \ 0.3 \ -0.2]^T, \ E_1 = [0.3 \ 0.2 \ 0 \ -0.5], \ E_2 = [0.4 \ 0.3 \ 0.6 \ 0.7].$$

The design parameter θ and ε are chosen as $\theta = 0.9$ and $\varepsilon = 0.05$. The state delay is $k_x = 1$.

7.5.1   General Case

The LMI (7.5) is solved using the LMI toolbox in MATLAB [16]. Based on Theorem 7.1, all the solutions are obtained simultaneously as follows:

$$Z = \begin{bmatrix} 1.166 & -0.091 & -0.169 \\ -0.091 & 1.135 & -0.108 \\ -0.169 & -0.108 & 1.234 \end{bmatrix}, Z_d = \begin{bmatrix} 1.794 & -0.174 & -0.294 \\ -0.174 & 1.680 & -0.134 \\ -0.294 & -0.134 & 1.678 \end{bmatrix}$$

$$Y = \begin{bmatrix} -0.009 & 0.019 & 0.239 \end{bmatrix},$$

and $\lambda = 1.726$.

Then the sliding gain matrix from (7.6) can be obtained as

$$K = \begin{bmatrix} 0.024 & 0.037 & 0.200 \end{bmatrix}.$$

The sliding function is calculated as

$$s(k) = \begin{bmatrix} 0.307 & 0.782 & 0.037 & 0.579 \end{bmatrix} x(k).$$

The control input is obtained as

$$u(k) = \begin{bmatrix} -0.207 & 0.072 & -0.049 & 0.224 \end{bmatrix} x(k)$$
$$+ \begin{bmatrix} -0.011 & -0.007 & -0.069 & 0.004 \end{bmatrix} x(k - k_x) - 0.034.$$

The matrices E and E_d are now obtained as

$$E = \begin{bmatrix} -0.207 & 0.072 & -0.049 \end{bmatrix}, E_d = \begin{bmatrix} -0.011 & -0.007 & -0.069 \end{bmatrix}.$$

The order of the observer is determined to be $q = 2$.

The matrices M, M_d, V, and V_d are chosen as

$$M = \begin{bmatrix} 0.1 & 0 \\ 0 & 0.4 \end{bmatrix}, M_d = \begin{bmatrix} 0.3 & 0 \\ 0 & 0.2 \end{bmatrix}, V = \begin{bmatrix} 1 & 2 \end{bmatrix}, V_d = \begin{bmatrix} 5 & 9 \end{bmatrix}.$$

The matrix $\mathcal{X}$ is now obtained as $\mathcal{X} = [\mathcal{X}_1 \quad \mathcal{X}_2]$, where

$$\mathcal{X}_1 = \begin{bmatrix} -0.143 & 0.056 & -0.114 & 0.149 & -0.151 \\ -0.107 & 0.002 & 0.048 & 0.003 & 0.076 \end{bmatrix}$$

and

$$\mathcal{X}_2 = \begin{bmatrix} 0.109 & -1.121 & 0.197 & 0.011 & 0.381 \\ -0.024 & 0.559 & -0.099 & -0.006 & -0.084 \end{bmatrix}.$$

Solving (6.37) and (6.38) gives the matrices J and J_d,

$$J = \begin{bmatrix} -0.143 & 0.056 & -0.114 \\ -0.107 & 0.002 & 0.048 \end{bmatrix}, J_d = \begin{bmatrix} 0.149 & -0.151 & 0.109 \\ 0.003 & 0.076 & -0.024 \end{bmatrix},$$

and solving (6.39) for the matrix T gives

$$T = \begin{bmatrix} -1.122 & 0.197 & 0.011 & 0.381 \\ 0.559 & -0.099 & -0.006 & -0.084 \end{bmatrix}.$$

Since $H = TB$, the matrix H is obtained as

$$H = \begin{bmatrix} -0.347 \\ 0.277 \end{bmatrix}.$$

Thus an estimate of $u(k)$ is given by the functional observer (6.25).

The resultant plots are shown in Figs. 7.1, 7.2 and 7.3. Figure 7.1 shows the response of the control input $u(k)$. Figure 7.2 shows the evolution of the estimated error $e_u(k)$. Note that the estimation error converges to zero. Figure 7.3 shows the sliding function response $s(k)$. The sliding surface variables converge to a neighbourhood of the sliding band, which verifies that the proposed DSMC ensures the existence of the sliding mode. The sliding function $s(k)$ lies within the band defined by $W_c = 0.556$.

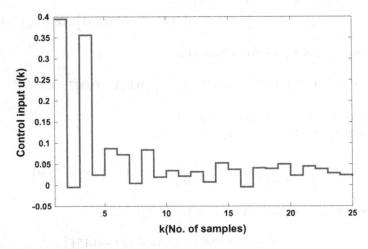

Fig. 7.1 Evolution of the control input $u(k)$

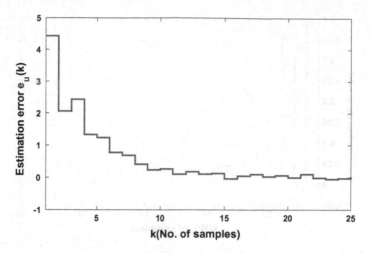

Fig. 7.2 Evolution of the estimation error $e_u(k)$

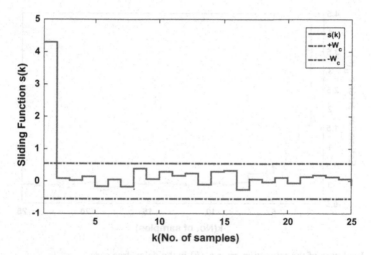

Fig. 7.3 Evolution of the sliding function $s(k)$

7.5.2   *Internal Delay-Free Observer Case*

Using a similar operation as in the general case, the structure of the functional observer form without internal delay can be expressed as

$$E = \begin{bmatrix} -0.207 \ 0.072 \ -0.049 \end{bmatrix}, \ E_d = \begin{bmatrix} -0.010 \ -0.007 \ -0.069 \end{bmatrix}.$$

The matrices M, V, and V_d will be the same as in the previous case:

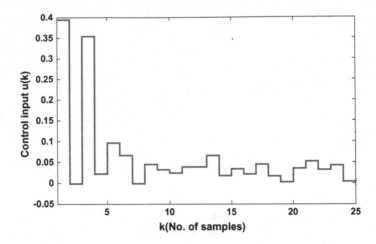

Fig. 7.4 Evolution of the control input $u(k)$ in the delay-free case

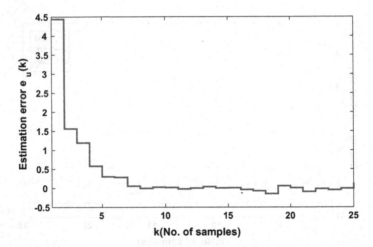

Fig. 7.5 Evolution of the estimation error $e_u(k)$ in the delay-free case

$$J = \begin{bmatrix} -0.302 & 0.111 & -0.144 \\ 0.040 & -0.084 & 0.055 \end{bmatrix}, J_d = \begin{bmatrix} 0.005 & -0.004 & 0.161 \\ 0.019 & 0.013 & -0.047 \end{bmatrix}, H = \begin{bmatrix} -0.521 \\ 0.372 \end{bmatrix}.$$

Thus the estimate of $u(k)$ is given by the functional observer (6.48).

Figure 7.4 shows the response of the DSMC $u(k)$. Figure 7.5 shows the evolution of the estimated error $e_u(k)$. Figure 7.6 shows the sliding function response $s(k)$. The sliding function $s(k)$ lies between the band defined by $W_c = 0.5465$. These simulation results demonstrate that the proposed design is very effective.

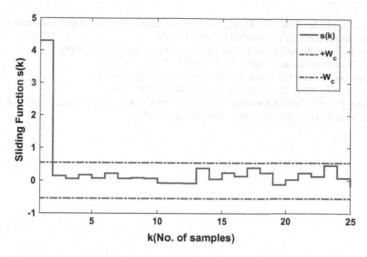

Fig. 7.6 Evolution of the sliding function $s(k)$ in the delay-free case

7.6 Conclusions

A sliding function and controller have been designed for a parametric uncertain time-delayed stochastic system. Further, a functional observer-based SMC for the time-delayed stochastic system was designed in the presence of parametric uncertainty in each of the system matrices. On the other hand, to mitigate the side effect of the parametric uncertainty on the dynamics of the estimation error, a sufficient condition on stability was given by Gershgorin circle theorem. Finally, the obtained theoretical results were validated by simulation results.

References

1. Hu, J., Wang, Z., Gao, H., Stergioulas, L.K.: IEEE Trans. Ind. Electron. **59**(7), 3008 (2012). https://doi.org/10.1109/TIE.2011.2168791
2. Singh, S., Janardhanan, S.: Int. J. Syst. Sci. **49**(9), 1895 (2018). https://doi.org/10.1080/00207721.2018.1479463
3. Boukal, Y., Zasadzinski, M., Darouach, M., Radhy, N.E.: In: American Control Conference, ACC'2016. United States, Boston (2016)
4. Islam, S.I., Lim, C.C., Shi, P.: J. Frankl. Inst. (2018)
5. Lin, Y., Shi, Y., Burton, R.: IEEE/ASME Trans. Mechatron. **18**(1), 1 (2013). https://doi.org/10.1109/TMECH.2011.2160959
6. Nguyen, M.C., Trinh, H., Nam, P.T.: Int. J. Syst. Sci. **47**(13), 3193 (2016)
7. Ji, Y., Qiu, J.: Appl. Math. Comput. **238**, 70 (2014)
8. Pai, M.C.: Proc. Inst. Mech. Engineers. Part I J. Syst. Control. Eng. **226**(7), 927 (2012). https://doi.org/10.1177/0959651812445248
9. Xia, Y., Fu, M., Shi, P., Wang, M.: IET Control Theory Appl. **4**(4), 613 (2010)
10. Xia, Y., Jia, Y.: IEEE Trans. Autom. Control **48**(6), 1086 (2003)

11. Darouach, M.: In: 2006 14th Mediterranean Conference on Control and Automation, pp. 1–5 (2006). https://doi.org/10.1109/MED.2006.328735
12. Singh, S., Janardhanan, S.: IET Control Theory Appl. **13**(4), 562 (2019)
13. Singh, S., Janardhanan, S.: Int. J. Syst. Sci. **48**(15), 3246 (2017)
14. Bartoszewicz, A.: IEEE Trans. Ind. Electron. **45**(4), 633 (1998)
15. Darouach, M.: IEEE Trans. Autom. Control **50**(2), 228 (2005)
16. Gahinet, P.: LMI Control Toolbox: For Use with MATLAB; User's Guide; Version 1. Computation, visualization, programming (MathWorks) (1995)

Index

© Springer Nature Switzerland AG 2020
S. Singh and S. Janardhanan, *Discrete-Time Stochastic Sliding Mode Control Using
Functional Observation*, Lecture Notes in Control and Information Sciences 483,
https://doi.org/10.1007/978-3-030-32800-9

113

Printed in the United States
By Bookmasters